AF773545

Biological Naturalism and the Mind-Body Problem

Jane Anderson

Biological Naturalism and the Mind-Body Problem

palgrave
macmillan

Jane Anderson
Department of Philosophy
University of Johannesburg
Johannesburg, South Africa

ISBN 978-3-030-99683-3 ISBN 978-3-030-99684-0 (eBook)
https://doi.org/10.1007/978-3-030-99684-0

© The Editor(s) (if applicable) and The Author(s), under exclusive licence to Springer Nature Switzerland AG 2022
This work is subject to copyright. All rights are solely and exclusively licensed by the Publisher, whether the whole or part of the material is concerned, specifically the rights of translation, reprinting, reuse of illustrations, recitation, broadcasting, reproduction on microfilms or in any other physical way, and transmission or information storage and retrieval, electronic adaptation, computer software, or by similar or dissimilar methodology now known or hereafter developed.
The use of general descriptive names, registered names, trademarks, service marks, etc. in this publication does not imply, even in the absence of a specific statement, that such names are exempt from the relevant protective laws and regulations and therefore free for general use.
The publisher, the authors and the editors are safe to assume that the advice and information in this book are believed to be true and accurate at the date of publication. Neither the publisher nor the authors or the editors give a warranty, expressed or implied, with respect to the material contained herein or for any errors or omissions that may have been made. The publisher remains neutral with regard to jurisdictional claims in published maps and institutional affiliations.

This Palgrave Macmillan imprint is published by the registered company Springer Nature Switzerland AG.
The registered company address is: Gewerbestrasse 11, 6330 Cham, Switzerland

Contents

1

Introduction

Human conscious awareness is undeniably one of the most incredible, and most baffling, of all our capacities. On the one hand, we know that our introspective conscious awareness comes about as a result of biological processes taking place in the brain; but, on the other, we have little understanding of how some biological processes get to have a conscious introspective feel to them, while others do not. Our bodies digest food, protect themselves against infection, regulate their own temperatures; yet none of these biological processes is experienced consciously. What is it that makes the biological process of introspective conscious awareness different from 'unconscious' biological processes like digestion, immune function, and temperature regulation? This is 'the mind-body problem': the 'problem' of how to find a place for conscious awareness within a physical world.[1]

In 1952, Roger Sperry wrote that "[d]espite steady advancement in our knowledge of the brain, the intrinsic nature of mind and its relation to cerebral excitation remains as much an enigma today as it was a hundred years ago" (291). In 1974, Thomas Nagel wrote that "the usual examples [of reduction] do not help us to understand the relation between mind and body" (436); not only do we not know how 'mind' is related to

© The Author(s), under exclusive license to Springer Nature Switzerland AG 2022

J. Anderson, *Biological Naturalism and the Mind-Body Problem*,
https://doi.org/10.1007/978-3-030-99684-0_1

'cerebral excitation,' but "we do not even have a clear idea of how the brain *could* work to cause consciousness" (Searle 1997: 201). More than fifty years after Sperry, Antonio Damasio wrote that "it would be foolish to presume definitive answers" to questions like "how does the brain construct a mind?" and "how does the brain make that mind conscious?" (2010, chpt. 1, sect. 1, para. 1). It is no longer the case that we have very little understanding of the brain and its workings (or at least, we understand it much better now than Sperry did in 1952)—but none of our knowledge about the brain seems to get us any closer to understanding 'how the brain constructs a mind.'

But before we have any hope of being able to answer a certain question, we need to be sure that we are asking the correct question. That is why a large part of this book is devoted to interrogating different ways of approaching 'the mind-body problem,' and showing why they are unsatisfactory. Towards the end, I will give some slightly more definite suggestions about what I consider to be the correct way to set about making sense of the problem—but at no point will I suggest that I have 'solved' the problem.

'The mind-body problem' is also sometimes referred to as the problem of 'the explanatory gap' between mind and body. That is what Chap. 2 is about. In Sect. 2.2, I examine the two main metaphysical frameworks within which people have tried to make sense of the explanatory gap: dualism and monism. Strictly speaking, monist positions can be either materialist or idealist, but idealism is hardly ever taken seriously in our current intellectual climate—so to all intents and purposes, the two candidates are dualism and materialism. Seeing that these two positions are usually taken to be mutually exclusive and jointly exhaustive, it is very difficult for most philosophers to conceive of a world which neither is dualist nor materialist—but that is exactly what I claim in this section: both dualism and materialism are deeply problematic, and we should reject them both. If neither dualism nor materialism enables us to account for 'the mind' (or 'thinking' or 'consciousness') in physical terms, and if these two positions are mutually exclusive and jointly exhaustive, then it follows, I argue, that it is not possible to produce a coherent metaphysical account of the explanatory gap.

My first claim—namely, that we lack a coherent metaphysical understanding of the explanatory gap—is not new (for example, Searle 1997 was already claiming that we should reject both dualism and materialism); but where I have gone further than people like Searle is that I claim that anybody who even articulates sentences containing terms like 'mind' and 'body' (let alone 'the mind-body problem') must actually be a 'closet dualist.' David Papineau (2011) has made virtually the same claim (and 'closet dualist' is his term), but this view has not been very generally adopted. Moreover, neither Papineau nor anybody else (as far as I know) has said anything similar to my further claim (2.2.3.1) that *the vicious reification of 'mind'* is a reliable indicator of 'closet dualism.' I argue that the very act of using 'the mind' in a sentence constitutes a tacit endorsement of dualism. Any view according to which there is an 'explanatory gap' between 'the mind' and 'the body' must be some variety of dualist view—and in this book, I will only be considering naturalist (monistic) views as satisfactory.

In this book, I use scare quotes to approximate the phrase 'so-called.' I argue that it is impossible to accommodate the notion of 'the (so-called) mind' (and/or so-called 'consciousness') within either a dualist or a materialist/naturalist framework—which amounts to much the same thing as saying that 'the (so-called) mind' (and/or so-called 'consciousness') *does not exist*. Another way of putting this, which I will utilize in Sect. 2.3.1.3.3, is to say that 'the [so-called] mind' exists only *in language*, and not *in reality*.

In Sect. 2.3, I examine some specific philosophical views relevant to the explanatory gap. I find that it is very difficult, if not impossible, to produce a coherent conceptual (philosophical) account of the explanatory gap. In fact, I claim, the conceptual (philosophical) accounts of the explanatory gap are problematic for *exactly the same reason* that dualist and materialist accounts are problematic: by continuing to employ the concepts 'mental' and 'physical,' these philosophers inadvertently demonstrate that they are actually closet dualists. Therefore, to the extent that neither dualism nor materialism can provide a coherent account of an 'explanatory gap'; to the extent that dualism and materialism are mutually exclusive and jointly exhaustive; and especially to the extent that almost all philosophical accounts are either overtly dualist or 'closet' dualist—I argue that it is also not possible to produce a satisfactory conceptual (philosophical) account of this gap.

Again, any number of writers have raised complaints about the kinds of philosophical accounts I discuss; but nobody, as far as I am aware, has complained that they are guilty of the vicious reification of 'mind'—even less that these accounts, *by reason of vicious reification*, can be described as 'closet dualist' accounts.

Given that I have rejected these typical metaphysical and conceptual (philosophical) accounts of the explanatory gap, the rest of the book is devoted to searching for a satisfactory alternative account. In Sect. 3.2, I consider some typical 'consciousness' research, and discover that much of this turns out to be research into what David Chalmers (1995) calls 'the easy problems of consciousness.' Whereas Chalmers uses the term 'easy problems (of consciousness)' to refer to those problems that 'seem directly susceptible to the standard methods of cognitive science,' my own view is that almost all of Chalmers' 'easy problems' could better be characterized as 'problems to do with ('conscious') perception and cognition.'

But taking questions about different kinds of ('conscious') perception and cognition to be 'the easy problems *of consciousness*' is, as I will show, problematic. I will demonstrate that much of our perception and cognition is in fact *not* 'conscious.' The implication is that if at least some perception and cognition is not 'conscious' (and especially if *much*, or even *most*, perception and cognition is not 'conscious'), then we should be careful not to conflate *'consciousness'* with ('conscious') *perception* or *cognition*—in spite of the fact that this is precisely what so much so-called 'consciousness' research has done.

In the second part of the chapter [Sect. 3.3], I focus specifically on investigations into what Chalmers calls 'the hard problem' of phenomenal experience: what it is like, from a subjective point of view, to have conscious experiences. However, I determine that even those authors who do not completely ignore this hard problem very often turn out to be interested in (the hard problem of) what it is like *to perceive* things, or what it is like *to perform certain cognitive operations*. This is problematic, because much perception and cognition is unconscious; and there is *nothing*-it-is-like to perceive things (or perform cognitive operations) unconsciously. Therefore, I argue, we need to be just as cautious of conflating (hard-problem) *what-it-is-like/experience* with (perceptual-cognitive) *consciousness* [Sect. 3.3] as we need to be of conflating *consciousness* with (easy-problem) 'conscious' *perception* or *cognition* [Sect. 3.2].

It is not at all clear to me why the conflation of (hard-problem or easy-problem) 'consciousness' with perception and/or cognition is so widespread—after all, even in 1995, Chalmers had pointed out the problem, and several of the twentieth century writers I discuss in this section also recognize it. Perhaps it is just, as Chalmers said, because 'the hard problem' *is* so hard. From the outset, people struggle to know how even to approach 'the hard problem,' which is why they so often end up approaching one of the easy problems instead.

In Sect. 4.2, I examine Freudian ('psychodynamic') theory. Freud claims that most people who have had 'a philosophical education' take 'consciousness' to be equivalent to 'introspective awareness.' But this is problematic, seeing that, just like perceptual and cognitive processes, many important psychological processes can take place without being (introspectively) 'conscious.' Whereas there is *nothing*-it-is-like to carry out perceptual or cognitive functions 'unconsciously,' in the case of psychological processes, saying that a certain process is (introspectively) 'unconscious' in no way implies that there is nothing-it-is-like to experience that process. In fact, on the Freudian view, it is taken for granted that some of the most significant factors contributing to (psychological) what-it-is-like/experience are not available to introspection. If, as I argue, Freud's view succeeds in demonstrating this, then we must recognize how important it is not to conflate '(psychological) what-it-is-like/experience' with '(introspective) consciousness.'

The idea that the kinds of perceptual-cognitive processing I discussed in Chap. 3 can be unconscious is reasonably uncontroversial, but that kind of unconscious (perceptual-cognitive) processing has little or nothing to do with the Freudian kind of unconscious *psychological* processes. Although there has been some contemporary research into such processes, this area is still under-researched. Moreover, with very few exceptions, it seems to me, Freud's critics exhibit a strong tendency—as Freud himself had occasion to note [Sect. 4.2.1]—to criticize Freudian theory without having more than a very limited understanding of what that theory actually says. Therefore, my careful and close reading of Freud's work has intrinsic value insofar as it may help to rectify some of the prevalent misreadings of Freud put forward by his critics. Finally, virtually nothing has been written by philosophers about how Freud's view helps make sense of the dissociation of (psychological) what-it-is-like/experience from 'consciousness' (for example, Levine 2000: 1; Talvitie 2015: 163); but there is a small number of recent writers in both psychology and philosophy who share my view that what Freud says about unconscious psychological processes is both meticulously argued and intuitively persuasive.

Another common way of approaching 'the mind-body problem' is to take 'the mind' (or 'consciousness') as essentially synonymous with 'rationality.' Essential rationality is a necessary premise of all 'computational' accounts, which have until very recently enjoyed significant popularity within the mind sciences. But if all 'conscious' individuals are also all 'rational,' making sense of 'irrationality' in such individuals is difficult, if not impossible. In Sect. 4.3, I briefly outline how philosophical accounts typically fail to make sense of irrationality; I then discuss how the psychodynamic account—which clearly differentiates 'consciousness,' '(psychological) what-it-is-like/experience' and 'rationality' from one another—performs much better in this regard. (This more successful psychodynamic account of irrational behavior is closely linked to the psychodynamic account of un-/conscious psychological material outlined in Sect. 4.2.)

Anyone who is familiar with Freud's work would be well aware that the psychodynamic account does not take 'consciousness' or 'rationality' to be essential attributes of persons—but as far as I know, not many philosophers ever think about the extent to which the assumption that persons are both essentially (introspectively) conscious and essentially rational informs their own views. Therefore, if they think about Freud's views at all, they typically do not interrogate the tension between their own assumption—that persons are essentially (introspectively) conscious and rational—and Freud's assumption, that they are essentially neither. Because (on the Freudian view) the majority of psychological material is opaque to introspective awareness, it becomes possible for a single individual to be simultaneously motivated by logically incompatible motivations—whereas, I argue, this is what any philosopher who takes 'mental' to entail 'conscious' is powerless to explain.

Having discovered a number of problems with a number of frameworks within which people typically attempt to account for what-it-is-like/experience (in Chaps. 2, 3 and 4), in Chap. 5 I set out some of what we know about how brains function—which may also constrain the search for a satisfactory account. Then I introduce the condition *hydranencephaly* [Sect. 5.2]. Hydranencephalic children are born with most or all of their cerebral cortex damaged or absent, but they are nevertheless 'conscious.' The discussion of this condition thus allows me to show why the prevailing corticocentric view of 'consciousness' is problematic—which means that we should not take the mind-body-self to be dependent on the functioning of the cerebral cortex.

The idea that the mind-body-self is not dependent on the functioning of the cerebral cortex is generally accepted among neurological investigators, but the corticocentric fallacy still lingers with some tenacity in other mind sciences. Among those philosophers who take any interest in empirical concerns to do with the actual brain, the corticocentric fallacy also lingers; but I show why this is problematic. Moreover, explaining about hydranencephaly enables me to refute another potential challenge to the Freudian framework I set out in Chap. 4. Mark Solms (who is an expert both in Freudian theory and in neuroscience) thinks that certain facts about hydranencephaly/the brain are incompatible with the Freudian framework—but I argue that that problem lies not in what Freud himself says, only in *Solms' reading of* what Freud says.

In Sect. 5.3 I investigate the implications of an anti-corticocentric view for our understanding of the essential nature of the mind-body-self. I examine five specific views which take a subcortical, non-perceptual/-cognitive, and non-introspective/-rational type of awareness to be foundational for all types of what-it-is-like/experience. This section provides the first part of my answer to the question of how to make sense of the explanatory gap: we *should* understand the (subcortical) mind-body-self—which is not essentially perceptual, cognitive, (introspectively) conscious, or rational—to be essentially affective. If we take this view, any cognitive or computational-type account becomes problematic.

As I mentioned, the anti-corticocentric view is widely accepted in neurological circles, but it severely challenges the many computational-inspired or cognitive views which still prevail widely in philosophy and the cognitive sciences. Moreover, neurological investigators seldom care about the theoretical, conceptual or philosophical implications of their findings—which is why there is a gap to be filled by a philosopher who takes on this challenge. Section 5.3 comprises my attempt to fill just such a gap.

Finally, in Chap. 6, I examine two specific views which qualify as examples of 'biological naturalism,' and thus suffer from none of the problems inherent in the 'dualist' or 'materialist' accounts I discussed in Chap. 2. In Sect. 6.2, I examine Antonio Damasio's 'body-map-based' view, focusing on how it performs in the respects I have discussed in the preceding chapters. Then in Sect. 6.3, I set out Jaak Panksepp's 'affect-centric' view, focusing especially on how this view compares with Damasio's view in these same respects.[2]

I have already explained why views which focus on 'conscious' perception and cognition (Chap. 3); views which conflate 'consciousness' either with introspection or rationality (Chap. 4); and corticocentric views (Chap. 5) are problematic from the point of view of anyone who is interested in what-it-is-like/experience. Therefore, demonstrating that Damasio's view is characterized by its emphasis on just these topics amounts to demonstrating that this view is problematic; and that, to the extent that it is opposed to the body-map-based view in these respects, the affect-centric view is therefore preferable. I argue that the affect-centric view is not only more satisfactory than any of the other views I have examined, in and of itself, but also (therefore) much more satisfactory with regard to making sense of the (illusion of the) explanatory gap.

The two views discussed in the final chapter (on 'Twenty-first-century biological naturalism')—but especially the 'affect-centric' view—enable us to escape from "the illusory sense that we are creatures of two distinct realms, of mind and matter" (Panksepp 1998: 302). Rather than talking about 'the explanatory gap' between 'the mind' and 'the body,' we *should* use a biological naturalist framework—something along the lines of the affect-centric view—to enable us to understand the mind-body-self.

Philosophers are not normally familiar with the psychological or biological empirical considerations constraining a theoretical framework for understanding the mind-body-self; psychologists are not normally concerned with either the biological considerations constraining, or the philosophical significance of, their empirical (psychological) findings; and biological researchers are not normally concerned either with the psychological or the broader philosophical implications of their empirical (biological) findings. Even more to the point, many researchers of all these varieties are not familiar with Freud. Of the writers whose work I examine in this book, Panksepp comes the closest to appreciating all these important matters—he is very knowledgeable about psychology (including Freud) and biology, and he even has some insightful things to say about philosophical concerns such as dualism in the neurosciences—but he does not have the necessary background to say anything much about how any of this should inform a philosophical understanding of the mind-body-self. In fact, there are very few writers who have the necessary expertise to draw the connections among all these areas. This is what I take to be both important and original about my project: the integration of all these lines of enquiry in the pursuit of a comprehensive new mind-body-self paradigm.

Notes

1. This introductory paragraph is reproduced in my article 'Damasio's body-map-based view, Panksepp's affect-centric view, and the evolutionary advantages of consciousness' (Anderson 2019: 419).
2. Portions of this chapter are also published in Anderson (2019).

References

Anderson, J. 2019. Damasio's body-map-based view, Panksepp's affect-centric view, and the evolutionary advantages of consciousness. *South African Journal of Philosophy* 38 (4): 419–432. https://doi.org/10.1080/02580136.2019.1697569.

Chalmers, D.J. 1995. Facing up to the problem of consciousness. http://consc.net/papers/facing.pdf. Accessed 23 March 2015. [Also/published in *Journal of Consciousness Studies* 2(3): 200–219)].

Damasio, A. 2010. *Self comes to mind: constructing the conscious brain*. New York: Pantheon books.

Freud--Complete Works, Ivan Smith. 2000, 2007, 2010. https://www.valas.fr/IMG/pdf/Freud_Complete_Works.pdf. Accessed 20 September 2015.

Levine, M.P. 2000. Introduction: How right does psychoanalysis have to be? In *The analytic Freud: Philosophy and psychoanalysis*, ed. Michael Levine, 1–7. London: Routledge.

Panksepp, J. 1998. *Affective neuroscience: The foundations of human and animal emotions*. ProQuest Ebook Central database: https://0-ebookcentral-proquest-com.ujlink.uj.ac.za/lib/ujlink-ebooks/detail.action?docID=431176#.

Papineau, D. 2011. What exactly is the explanatory gap? *Philosophia* 39: 5–19.

Searle, J.R. 1997. *The mystery of consciousness*. 7th ed. New York: NYREV.

Talvitie, V. 2015. Beyond the philosophy of the (unconscious) mind: The Freudian cornerstone as scientific theory, a cult, and a way of talking. In *Psychoanalysis & Philosophy of Mind: Unconscious Mentality in the 21st Century*, ed. S. Boag, L.A.W. Brakel, and V. Talvitie, 163–192. ProQuest Ebook Central database: https://0-ebookcentral-proquest-com.ujlink.uj.ac.za/lib/ujlink-ebooks/detail.action?docID=2082961&query=psychoanalysis+and+philosophy+of+mind. Accessed 10 September 2019.

2

The Explanatory Gap

2.1 Introduction: The Explanatory Gap and What-it-is-like

Identifying one concept with another is one of the most common ways of 'explaining' things. For instance, we propose that our ordinary language concept 'water' actually refers to the chemical compound H_2O; and we accept that talking about H_2O is exactly the same as talking about 'water.' There is nothing in our common-sense concept of water that is left out when we refer to it in the scientific terms ('H_2O'); and there is no part of the scientific concept of H_2O which is left out when we refer to 'water' in our everyday language (Papineau 2011). But in the case of human conscious awareness and the human brain, there appears to be no straightforward way in which to identify the one with the other. Even if we know exactly what happens in somebody's brain when that person is having a particular subjective (what-it-is-like) experience, "we have a strong sense that" knowing what happens in the brain does not fully *explain* the what-it-is-like/experience (Tye 1999: 705). "[W]e have at present no conception of what an explanation of the physical nature of a mental phenomenon would be" (Nagel 1974: 436)—never mind having this actual explanation itself.

© The Author(s), under exclusive license to Springer Nature Switzerland AG 2022

J. Anderson, *Biological Naturalism and the Mind-Body Problem*,
https://doi.org/10.1007/978-3-030-99684-0_2

In his 1983 article, Joseph Levine coined the phrase 'the explanatory gap' to refer to the idea that "we don't have any way of determining exactly which psycho-physical identity statements are true"—that is to say, there appears to be an unbridgeable 'gap' between psychological or mental phenomena, and physical phenomena. We can explain psychological phenomena; we can explain physical phenomena; but we cannot explain how some psychological phenomena and some physical phenomena are connected. Even if we are told (and even if we truly believe) that 'the mind' is identical to 'the brain,' describing brain processes seems to be inescapably different from describing 'mental' processes. For instance: we may be told that when we feel pain, what is really going on is that the C-fibres in our brains are firing. But knowing about what your C-fibres are doing will not be enough to tell you what that pain actually *feels like* (Papineau 2011: 6). The concept *pain*, in our language, has nothing to do with the physical processes that may or may not give rise to pain feelings: it refers to what it is like to *experience* the pain. Michael Tye says "[i]n the case of phenomenal consciousness ...we can still coherently ask why so-and-so brain processes or functional states feel the way pains do or why these processes feel any way at all" (1999: 706).

Thomas Nagel famously proposed that when an organism is having a subjective (phenomenal) experience, "there is something it is like to *be* that organism" (1974: 436); and I am going to follow Chalmers' lead in taking Nagel's kind of phenomenal experience ('what-it-is-like') as the crux of the mind-body problem. Nagel thought that "the problem of conscious experience is what makes the mind-body problem both interesting and intractable" (Montemayor and Haladjian 2015: 7)—and this problem of conscious experience (that is, 'the mind-body problem' and/or 'the problem of the explanatory gap') is what we are no closer to understanding, even today.

In Sect. 2.2, I examine the two main metaphysical frameworks within which people have tried to make sense of the explanatory gap between 'the mind' and 'the body': dualism and materialism. Our unquestioning acceptance of the assumption that either dualism is false and materialism is true, or that materialism is false and dualism is true (that is, that these positions are mutually exclusive and jointly exhaustive), infuses all our thinking about the explanatory gap; but I will argue that both dualism and materialism are unsatisfactory, both in and of themselves, and (in particular) with regard to the explanatory gap problem. What is more, I will argue, many of those who profess to be materialists actually betray their 'closet dualism' when they use the term 'the mind' in exactly the same way they use other terms, such as 'the body'—which is to say, when they commit the logical fallacy of *vicious reification* with regard to 'the mind.'

The only people who can talk about 'the mind-body problem' are people who believe that 'the mind' *exists* in the first place—and I argue that this belief is constitutive of dualism. It is not possible to 'solve' the mind-body problem (which is to say, the problem of the explanatory gap between mind and body) because it is not possible, within either dualism or materialism, to articulate coherently how physical and mental phenomena could relate to each other. In Sect. 2.3, I apply analysis of dualism and materialism to four specific examples of twentieth-century views in the philosophy of mind. I argue that—primarily in virtue of the fact that they commit vicious reification—these views do not contribute towards producing a coherent account of the explanatory gap.

2.2 Dualism, Materialism, and the Explanatory Gap

Monism is characterized by the proposition that the world is made of only one kind of stuff. With regard to the problem of the mind and the body (that is to say, of the explanatory gap), monist views are either Idealist (the view that only mind, or spirit, exists) or Materialist (the view that only matter, or the physical body, exists). *Materialism* has generally been the default position, owing to the hegemony of 'the scientific method': now that our modern techniques and instruments have given us so much insight into how physical bodies of all descriptions function, it is much more difficult to stick to any view that entails any kind of non-physical or supernatural causal entities.

But in spite of this, "[w]e find it almost impossible to free ourselves from the dualist thought that conscious feelings must be something *additional* to any material goings-on" (Papineau 2002: 3). *Dualism*, insofar as the explanatory gap is concerned, is the view that "mental states and processes are not merely states and processes of a purely physical system, but constitute a distinct kind of phenomenon that is essentially *non*physical in nature" (Churchland 2013: 3). However, most philosophers would agree with Searle that dualism is unsatisfactory: having made a strict distinction between the mental and the physical, dualism cannot then say anything about how 'mental' things and 'physical' things could relate to each other.

Our unquestioning acceptance of the assumption that either dualism is false and physicalism is true, or that physicalism is false and dualism is true (that is to say, that dualism and physicalism are mutually exclusive and jointly exhaustive), infuses all our thinking about the explanatory gap. But I am going to try and show why both dualism and materialism are unsatisfactory both in and of themselves, and (in particular) with regard to making sense of the explanatory gap. In Sect. 2.2.1, I am going to focus in some depth on the form of dualism articulated by Descartes, because Descartes' ideas underlie virtually all modern philosophical thinking. In the section on the explanatory gap and materialism [Sect. 2.2.2], I do not focus on any one particular writer.

2.2.1 The Explanatory Gap and (Cartesian) Dualism

Cartesian dualism is the view that 'the mind' is some kind of non-physical substance, related in some unspecified way to 'the physical body,' from which it is different in principle—as opposed to different only in detail. Whereas the essential property of matter is that is 'extended' in space, the essential property of the mind is that it thinks. Cartesian ideas "dominated the scientific and philosophical thinking of Europe" for a considerable time, owing to the clarity and precision with which these ideas were expressed; as well as the "remarkably wide range" (Cottingham 1996: xxxvii) of topics about which Descartes was knowledgeable. Because his views have been so influential, in a sense, my whole investigation of the 'explanatory gap' could be said to be devoted to interrogating the lingering influence of Descartes' view.

2.2.1.1 Bodies

In nineteenth-century neuroscience, the two prevailing metaphysical positions were (Cartesian) substance dualism, and epiphenomenalism (which was the view that 'the mind' existed but was causally redundant). The leading proponent of epiphenomenalism was Thomas Henry Huxley: Huxley's article 'On the Hypothesis that Animals are Automata was a keynote text of the Victorian period, which had a strong impact on late Victorian literature and culture. Huxley writes that in the early sixteenth century, the idea that vital phenomena, like physical phenomena, are capable of mechanical explanation started to take shape. William Harvey was the first to explain the mechanism of the circulation of the blood, which was a very important achievement; however, it is Descartes whom Huxley credits with having produced an account of the mechanism of animal feeling and movement, which even Harvey did not do. Huxley said that "[Descartes] was a working anatomist and physiologist, conversant with all the anatomical and physiological lore of his time, and practised in the methods by which anatomical and physiological discoveries were then made" (Huxley 1874: 362). In Huxley's opinion, Descartes was as conversant with anatomical theory and practice as any physician of that time; and he thought that Descartes' work incorporated some theories about the working of the nervous system which still provide the basis for modern physiology.

Most significantly, Descartes insisted that all physical structures, including non-human creatures, could be accounted for exclusively in terms of the functioning of their physical components. The basis for Descartes' view that animals function essentially automatically was the observation that many of our actions occur 'reflexively,' and without volition. Because reflex actions, and even some much more complex actions, are not even conscious—let alone voluntary—it is clear that 'consciousness' is not required for the essential functions of maintaining biological life. Humans, on Descartes' view, differ from other creatures in virtue of their having divine souls; but he accepted that quite possibly "the whole of man's physical actions are mechanical, his mind living apart … and only occasionally interfering by means of volition" (Descartes, in Huxley 1874: 364).

Because Descartes believed that thinking things and physical things are different in principle, it is natural to assume that he did not understand how important the brain was for the production of thinking/consciousness. Cottingham, for instance, notices that Descartes says: "if a foot or arm or any other part of the body is cut off, nothing has thereby been taken away from the mind" (1641/1996: 59). From this, Cottingham suggests, it follows that "the *brain*, being a purely bodily organ, must, for Descartes, be inessential to the mind's continued functioning as foot or arm" (Cottingham 1996: xxxi)—but Cottingham's inference is not justified:

Descartes does say that "the mind is not immediately affected by all parts of the body," and this is why it is possible to cut off a foot or arm without affecting 'the mind'; but he also notes that in the case of the brain (or rather "perhaps just … one small part of the brain"), physical changes *do* 'affect' the mind. He proceeds to suggest that this one small part of the brain must be whichever part contains the 'common' sense (which he understood as "the supposed faculty which integrates the data from the five specialized senses," 1641/1996: 59)—which at that point in history was assumed to be the pineal gland. But although we are now confident that the pineal gland does not perform this kind of 'integratory' function, we can also be confident that Descartes, as a physician, knew perfectly well that the brain was essential for his thinking thing.

Cottingham is therefore not justified in ascribing to Descartes the belief that the brain is as inessential to the mind's functioning as a foot or arm. Descartes' work reflects his conviction that "the brain is the organ of sensation, of thought, and of emotion" (Huxley 1874: 362). Although this is universally accepted today,

> in Descartes' time, and I may say for 150 years afterwards, the best physiologists had not reached that point. It remained … a question whether the passions were or were not located in the abdominal viscera. (Huxley 1874: 362)

However, what is also completely clear is that even if, on the Cartesian view, some parts of 'the body' have some kind of power to influence 'the mind,' mind and body are comfortably distinct from each other. They are different not only in detail, but also different in principle. Because he believed that thinking/consciousness was different in principle from physical structures, Descartes' view did not enable him to make sense of the explanatory gap—and neither, I will claim [Sect. 2.2.1.3], do more modern versions of dualism.

2.2.1.2 Minds

Descartes' famous *cogito*—I think, therefore I am—is very well-known among philosophers in all branches of philosophy; but Hennig (2010: 7) suggests that Descartes himself may not actually have endorsed what we (now) refer to as Cartesianism (or at least, that his view was more subtle than people often give him credit for). It is conventional to take 'thinking' as an introspective kind of awareness, but this modern idea of 'thinking' is probably not what Descartes had in mind; he was probably referring to a kind of involuntary, pre-reflective awareness, rather than the conscious, intentional kind of thinking that contemporary philosophers tend to be interested in. Hill argues that when Descartes claimed that being human was characterized by 'thinking,' he was almost certainly using the term 'thinking' "in something like its ordinary, pre-philosophical meaning," as "an intellectual power involving judgment" (2010a: 3). For this reason, Hill believes that, contrary to the dominant tradition, we should take Descartes' use of the term 'thinking' in its more intellectual sense, rather than in the broader sense of 'consciousness' (2010b: 41).

On the Cartesian view, exercising their faculty of rational judgment is the characteristic activity of 'thinking things' such as human persons. Even what we normally take to be sense perception, for Descartes, turns out to be a question of rational judgment. Descartes claims that the only essential characteristic all physical objects share is 'extension' in space, and since extension can be inferred from the mere fact that an object exists, he concludes that physical objects *are not* in fact 'perceived' through the senses: they must be apprehended purely intellectually.

For example: when he looks out of the window, says Descartes, and appears to 'see' men passing by on the street, in reality the only thing he actually sees are their hats. From the appearance of the hats, he then *infers* that there are men walking along wearing the hats—but the hats could just as well conceal 'automatons' as actual men (1641/1996: 21). It might be practical to say that one 'sees' men passing by, when all that is really visible is their hats, but this is not completely accurate. Strictly speaking, all one *sees* is the hat, and then one *judges* that the presence of the hat indicates the presence of a man, wearing it. Similarly, in the case of the piece of wax which changes its form and appearance as it melts, "[t]he meditator finds that in order to perceive the piece of wax *as a piece of wax* through all the changes that it undergoes as it melts, he must be judging intellectually" (Hill 2010b: 40).

Descartes is not making an empirical claim, but rather stating a logical truth: even when it appears that one is perceiving the external world, what is really going on is that one is exercising the faculty of rational judgment which (for Descartes) epitomizes the mind. Therefore, on Descartes' view, thinking/consciousness can not only not be accounted for in terms of *the functioning of* physical structures [Sect. 2.2.1.1]; it can also not be accounted for in terms of *the sensory perception of* physical structures. For Descartes, physical structures *are not* perceived through the senses, but grasped by the faculty of understanding. That seems reasonable.

Although Descartes takes thinking things and their physical bodies to be "mutually independent substances, each of which can exist without the other," he also stresses the fact that they are very closely 'united' or 'intermingled' (Cottingham 1996: xxxiii). He does not appear to have a clear idea of how this might work, in practice; and in fact he acknowledged in a letter that the question of "how can the soul, a thinking substance, causally influence the behaviour of the body to bring about voluntary actions?" (Cottingham 1996: xxxv) was the one question he had not managed to answer.[1] The body we understand in terms of its being extended; the soul we understand in terms of thought (including perception and the will); but when the soul and body are 'united,' "we have only the notion of their union, on which depends our notions of the soul's power to move the body, and the body's power to act on the soul and cause its sensations and passions" (ibid.).

2.2.1.3 Properties and Substances

Most thinkers in the various 'mind sciences' now accept that extreme Cartesian dualism is not plausible, and that thinking/consciousness has some kind of relation to brain events (I will discuss a number of these thinkers in Chaps. 4 and 5); but what we still lack is a satisfactory understanding of how exactly this relation works. For example, contemporary researchers may suggest that "mental events can have *aspects* or *properties* which are phenomenal, subjective, or private" (Pataki 2000: 26), while at the same time the events themselves are somehow normal and physical. This response is supposed to enable property dualists to say both that everything that exists is material, and that thinking/consciousness is not merely a figment of our collective imagination. However, it seems to me that this does little to answer the underlying question: exactly *how it can be* that 'physical' processes can have 'mental' aspects or properties.

Pataki's re-formulation of the question of how 'the mind' could relate to 'the body,' in terms of how it can be that physical events can have mental *aspects or properties*, is necessarily based on the assumption that 'physical processes' and 'mental processes' (or 'properties') are different in kind—as was Descartes' formulation of the question, in terms of how the physical brain could produce thinking/consciousness. In other words, it is difficult to see how a materialist (who is committed to the view that only physical bodies exist) could, without contradiction, even recognize the existence of such a thing as 'a non-physical property.' Strictly speaking, we would probably need to say that materialists can recognize neither the existence of non-physical properties *nor even the existence of physical properties*—since 'a property,' of any description, is a theoretical construct, rather than a physical body; and materialists are supposed to believe that only physical bodies exist. Therefore, it seems much more appropriate to call anybody who talks about mental substances, aspects, or properties, a dualist, rather than a materialist.

Likewise, it does no good to suggest that "we have distinctive first-person *ways of thinking* or *modes of presentation* of the neural events which realize our experience" (Pataki 2000: 26)—because if neural events are to 'realize' our experience, then 'neural events' and 'experience' must be distinct from each other. Even if we have 'distinctive first-person ways of thinking' about these neural events, or 'modes of presenting' these neural events, anybody who says that neural events 'realize' experience must necessarily believe that 'neural events' and 'experience' are not the same thing. If 'neural events' and 'experience' were the same thing, then the sentence would have to read 'neural events realize neural events,' or 'experience realizes experience,' which is not only not informative: it is not even true—since the only relation that can obtain between a thing and itself is *identity*. Therefore, strictly speaking, the only true sentences that can be articulated about the relation of identity are, for example, 'neural events *are* neural events,' or 'experience *is* experience'.[2]

Accordingly, trying to bridge the explanatory gap by appealing to (first-person) *ways of thinking* or *modes of presentation* of the neural events which 'realize' our experience is just as problematic as trying to bridge it by appealing to the (mental) *aspects or properties* of these neural events. What is more, it is problematic for exactly the same reason: neither *aspects*, *properties*, *ways of thinking* nor *modes of presentation* are physical objects, which is why, on a materialist view, none of these 'things' can be said to 'exist' (not being physical objects)—let alone interact causally with other things.

In other words: the distinction between 'property dualism' and 'substance dualism' is spurious. All dualism is characterized by its endorsement of the claim that 'mental' and 'physical' substances (and/or 'properties' and/or 'aspects' et cetera) are different in kind, rather than just in detail.

2.2.1.4 The Lesson (of 'The Explanatory Gap and Cartesian Dualism') for Philosophy

As I said, Cartesian dualism, having made a strict distinction between the mental and the physical, cannot then make the relation of the two intelligible [Sect. 2.2]. Cartesian dualists deny that mental phenomena can be accounted for purely in term of physical processes, which means that they do not need to produce a convincing explanation of how exactly 'the mind' relates to 'the body'; but that also means that their position is no help for explaining how mental phenomena can have physical effects, or in fact be anything other than totally irrelevant (and 'epiphenomenal'). In other words, dualist accounts of the explanatory gap are unsatisfactory—or perhaps we could even say that, strictly speaking, there is *no such thing* as a dualist *account* of the explanatory gap.

Although dualism in the extreme form espoused by Descartes is no longer popular, I suggested that the distinction between the (discredited) Cartesian form of dualism and the (accepted) property dualism was spurious. I am going [Sect. 2.2.2.2] to expand on my suggestion that it is much more difficult than one might think to escape completely from a Cartesian-type paradigm, and that even many professed opponents of dualism display alarmingly dualist leanings. As Papineau says, "[t]he feeling of an 'explanatory gap' arises only because we cannot stop ourselves thinking about the mind-brain relation in a dualist way" (2011: 5). Even if I had not determined that dualism was problematic, I have stipulated that, for the purposes of this book, I will take a 'biological naturalist' position; I will therefore be assuming that all dualist accounts are problematic.

2.2.2 The Explanatory Gap and Materialism

Because the acceptance of dualism seems to require the rejection of the scientific worldview, most people try their best to claim that the only reality that exists is physical reality, maintaining that 'the mental' is some kind of 'special' aspect of material reality, or somehow 'arises' from physical reality. In other words, *materialism* "evolved as a way of rejecting dualism, as a way of denying that there are 'mental' things or features apart from, and metaphysically different from, the rest of the 'physical' world" (Searle 1997: 210).

There is nothing like universal acceptance of exactly how 'materialism' should be defined, but I will argue that materialism is necessarily an eliminativist position. It is the position that says 'Everything that exists is material (or physical)—therefore, *'the mind' (or what-it-is-like/experience) does not exist.*' Materialists recognize the existence of the many things in the world that *appear* to be non-physical—'things' we might describe as 'psychological,' 'moral,' or 'social'—but they maintain that there is a correct way of describing even these items as 'material' in essence (Stoljar 2015, Introduction). Seeing that materialists deny that 'the mind' (in the sense of something that is different in kind from 'the body') even *exists*, obviously, their position does not enable them to make sense of the intuition that there is any kind of 'gap' between 'the mind' and 'the body.'

Later on (from Chap. 3), I will contrast Materialism with Physicalism, which I take as the endorsement of a different claim: 'Everything that exists is physical (or material)—therefore *'the mind' (or what-it-is-like/experience) is physical.*' In this book, I assume that physicalism is the only plausible metaphysical view.

2.2.2.1 Reductive Materialism

Conventionally, reductive materialism can take two forms: the type-identity theory, and the token-identity theory. On the type-identity theory, *types* of mental phenomena are taken to be identical to *types* of neural phenomena—for example, the 'type' of sensation we refer to by the term 'pain' is taken to correspond to a type of neural event (such as a firing pattern among certain C-fibres). The main problem with type-identity theory is (unsurprisingly!) that it turns out to be very difficult to decide whether neural phenomena belong to the same, or different, 'types.' Moreover, it seems that at least some 'mental processes' which we would want to count as belonging to the same 'type' are associated with physiological processes which we would *not* want to count as belonging to the same 'type' (van Riel and Van Gulick 2019, Sect. 3.2). As a result of this kind of objection, the type-identity theory has largely been abandoned.

On the token-identity theory, it does not matter whether any particular psychological 'type' can be equated with any particular physiological type: what matters is that each and every specific instance (*token*) of a 'mental' event is identical to a specific physiological event. The token-identity theory is opposed to any kind of dualism; it therefore qualifies as a 'reductionist' view.

2.2.2.2 Non-reductive Materialism

Because it of the difficulty of producing a coherent type-identity theory, many philosophers have been led "to seek an alternative way of reconciling the efficacy of mental and other special causes with the causal closure thesis" (Papineau 2020, Sect. 1.4).

> The general idea of this 'non-reductive physicalism' [which I am calling 'non-reductive *materialism*'] is to allow that instantiations of a given special property will always be grounded in or metaphysically determined by instantiations of physical properties, but that these 'realizing' physical properties might be different in different cases. So, for example, any being who thinks about the square root of two will do so in virtue of instantiating some physical properties, but these can be different physical properties in different cases. (Papineau 2020, Sect. 1.4)

But the question is: does this 'non-reductive physicalism' (or 'materialism,' as I am using the term) genuinely qualify as a 'physicalist' position? As Wiredu observes, "if one entity is different from another entity, then that entity has a certain characteristic which the other does not share" (1996: 200)—which also implies that if one physical entity is identical to another physical entity, then neither entity can have any 'property' which the other entity lacks. Identical (physical) entities cannot have differing 'properties,' and distinct (physical) entities cannot have identical 'properties.' This means that if 'non-reductive physicalism' is the view that different physical bodies can have the same non-physical 'properties,' then the claim of that view to be a physicalist view is spurious. According to Safro Kwame, "it is apparent, or it ought to be, that a 'non-reductive physicalism [or 'materialism']' is really not a kind of physicalism" (2004: 343), since "any kind of physicalism which rejects dualism must in some minimal sense, at least, be reductive" (2004: 345).

2.2.2.3 Awareness of Mental States

Moreover, any kind of (non-/reductive) materialism/physicalism faces a serious difficulty when it comes to making sense of our introspective awareness of our own mental processes. If mental processes were exactly the same as brain processes, then *being aware of* mental processes and *being aware of* brain processes should also be the same; but this is certainly not often—or maybe even *ever*—the case. We have no introspective awareness of the processes going on in our brains as we have introspective awareness of what we think of as 'mental' processes. We can easily say 'I feel happy' or 'I am worried that I may not finish my essay on time,' but we can never say 'My cingulate gyrus seems to be a bit overactive today' or 'My right hemisphere primary motor cortex is so fatigued, I fear it may start making careless mistakes.' We cannot make such statements at the present time, and it is also very difficult to see how we might acquire the ability to make such statements at any future time, no matter how much we might find out about how the brain works.

Searle thinks that even materialist philosophers must implicitly acknowledge that awareness of 'physical (brain) processes' is not identical to awareness of 'mental processes.' This makes it virtually impossible for them to articulate a coherent materialist position. According to Searle, because of this difficulty, materialist philosophers actually spend most of their time trying to "get rid of the mental," rather than genuinely trying to offer 'an analysis' of mental concepts:

> It sounds too implausible to say right out that pains and beliefs and desires don't exist, though some philosophers have said that. The more common materialist move is to say, yes, mental states really do exist, but are not something in addition to physical phenomena; rather they can be reduced to, and are forms of, physical states. (Searle 1997: 136–137)

That is to say, materialists have to try and describe 'mental' concepts, like pains, beliefs and desires, in terms which neither require the postulation of any kind of non-physical substance, nor rule out the possibility of the existence of those very mental concepts they are supposed to be describing—which is either a very difficult, or a conceptually impossible, thing to do. Things which do not exist in the normal, physical sense are, generally speaking, not the kind of things which can be recognized on a twenty-first-century (physicalist) scientific worldview; which is why, according to Searle, saying that mental states do exist, *but are not something in addition to physical phenomena*, is tantamount to saying that mental states (or 'the mind') *do not* exist.[3] In which case: it appears that the distinction between reductive and non-reductive materialism may be just as spurious as the distinction between substance dualism and property dualism [Sect. 2.2.1.3].

2.2.2.4 The Lesson (of 'The Explanatory Gap and Materialism') for Philosophy

Accordingly, not only is the distinction between property dualism and substance dualism spurious [Sect. 2.2.1.3]; the distinction between reductive and non-reductive materialism is also spurious [Sect. 2.2.2.3]. I said that, if a so-called property dualist tried to bridge the explanatory gap by opposing 'physical properties' to 'mental properties,' it was difficult to see how s/he could justifiably maintain that s/he was not a substance dualist [Sect. 1.2.1.3]. Similarly, if a so-called (non-reductive) materialist tried to bridge the explanatory gap by opposing something like 'experience' to 'the neural events which realize our experience,' it is difficult to see how s/he could justifiably claim to believe that everything that exists is material [Sect. 1.2.2.3]—in other words, to maintain that s/he was not also a (substance) dualist. In fact, it is highly implausible for anyone who believes that both 'neural events' and 'experience' *exist*—in the same literal sense—to claim to be a materialist of any description.

Accordingly, materialist accounts of the explanatory gap are unsatisfactory—or perhaps we could even say that there is *no such thing as* a materialist account of the explanatory gap (more on this in 'Reification' below).

2.2.3 Dualism, Materialism, and the Explanatory Gap: Conclusion

Descartes suggested that thinking/consciousness had something to do with an 'immaterial substance'; more sophisticated (recent) forms of dualism usually do not go quite so far as this; but what is common to all of them is that they claim that brute physical phenomena (or *the aspects* or *properties of* brute physical phenomena) are different in kind from (the aspects or properties of) 'mental' or subjective phenomena, and that is why different kinds of laws apply to the two different classes of phenomena. This being the case, dualism is not competent to make any sense of the explanatory gap.

Materialists maintain that what we might describe as 'the mind' is a natural part of the normal physical world, and not different in principle from physical bodies—and it can be argued that this claim is tantamount to the claim that 'the mind' *does not exist* (since 'the mind' is defined as different in principle from material things, and materialists believe that everything which exists is material). Seeing that, according to (reductive/ eliminative) materialism, 'the mind' does not exist, materialism is *a fortiori* not competent to make any sense of the explanatory gap (between 'the mind' and 'the body').

Therefore, because neither dualism nor materialism is competent to make sense of the explanatory gap—and because dualism and materialism (in the limited sense in which I am considering them, for the purpose of this book) are mutually exclusive and jointly exhaustive—it is not possible to produce a coherent metaphysical account of the explanatory gap. I have suggested that the only people who are able to talk about 'the explanatory gap' *at all* are dualists—and I have stipulated [Sect. 2.2.1.4] that I will not consider dualism as a plausible view.

> [F]rom a purely logical … point of view, a theory of mind need not be a theory of the mind-body problem.
>
> Further, if a theory is to be illuminating, … one of its objectives should be to eliminate eliminable problems, not to create them. It would seem … that there is much to say in favor of … a theory of mind which eliminates

> the mind-body problem rather than raises it, provided, of course, that the elimination is achieved by sound methods. (Kwame 2004: 349)

2.2.3.1 Reification: Recognition of the Existence of an 'Explanatory Gap' is Constitutive of Dualism

It seems to me that the main thing that goes wrong with materialist accounts of the explanatory gap is that many of those who profess to be materialists repeatedly commit *vicious reification* as they attempt to expound their views. Reification, from the Latin *res* ('thing'), is the consideration or representation of an abstract concept as physical or concrete (in other words, treating a 'non-thing' as a 'thing'). In everyday usage, reification is a natural, innocuous linguistic device, and does no harm at all; but the use of reification in logical reasoning is misleading and usually regarded as a fallacy. This 'fallacy of ambiguity' is also known as 'concretism' or 'the fallacy of misplaced concreteness,' since it refers to the (inappropriate) treatment of an abstract concept as if it were a concrete entity. It is a fallacy because "[a] concept is, by definition, a non-object, and… an object is, by definition, a non-concept" (Wiredu 1987: 159).

An example of innocent, useful reification is the concept 'hurricane.' Saying that that *the hurricane* caused damage in certain parts of the country is really a shorthand way of saying that *a certain group of identifiable meteorological phenomena* caused damage in those regions—which is why this reification is 'innocent.' There is nothing intrinsic about the various meteorological phenomena which comprise a hurricane that calls for their necessarily being grouped together in this definitional way; but, from the point of view of creatures who live above ground (as opposed at the bottom of the sea, for example), and fall within an approximate size range (hurricanes being meaningless both to bacteria and to solar systems), using the term 'hurricane' is convenient and harmless. We routinely use innocent reification when we talk about things like *pairs of* gloves (as opposed to numbers of individual gloves), or when we refer to *the Auckland Park campus* of the university (as opposed to the individual buildings which comprise this campus).

Similarly, if *the mind* is understood to signify simply (the experiential aspect of) 'a group of biological/neurological phenomena,' then it is an example of innocent reification, and is acceptable for the dualist and the materialist alike. If saying that 'I made up my mind to get out of bed' is taken to be just a kind of figurative (or 'shorthand') way of saying 'a group of biological/neurological phenomena caused the relocation of a certain physical body,' then the reification involved is innocent. But if the situation is understood as a case in which 'the mind' *causes* the relocation of the physical body, then the reification is vicious, and constitutes a fallacy.

Compare this with, for example, the way we refer to hurricanes. If the term 'hurricane' is used as a convenient shorthand to refer to the occurrence of certain weather patterns, then it is an instance of innocent reification. On the other hand, as soon as somebody articulates a sentence about 'the relationship' between 'the hurricane' and 'the meteorological phenomena,' this indicates that that person considers 'the hurricane' to be some kind of actual (causal) entity—as opposed to a mere linguistic convenience. The same applies in the case of 'the mind': if somebody *reifies* 'mind'—even if that person proceeds to say something like 'the mind *just is* the brain'--the fact that the person uses different terms to refer to 'the mind' and 'the brain' shows that, on her understanding, it is (in at least some sense) meaningful to distinguish between the two concepts.[4]

Wiredu uses the term 'hypostasis,' rather than the term '[vicious] reification,' but he seems to be thinking along much the same lines. On his view, to refer to 'the mind' in the same way that one refers to physical objects "is to trifle with … the most basic ontological distinction in the whole gamut of human thinking, namely, the distinction between concept and object" (1987: 164).[5] In other words: presuming that 'mind' is an 'entity' is unacceptable not only for the materialist, but also to anybody who communicates using a natural language. Wiredu says that "to convert something abstract … into an object or entity is to commit the error of hypostasis" (1987: 172)—which seems very similar to my saying that 'to consider an abstract concept as physical or concrete is to commit the fallacy of reification.'

Papineau evidently agrees with this analysis:

> Brain processes are standardly said to 'generate', or 'yield', or 'cause', or 'give rise to' conscious states. … [But] [t]o speak of brain processes as 'generating' conscious states, and so on, only makes sense if you are implicitly thinking of the consciousness as ontologically additional to the brain states. (2011: 11–12)

The same applies to 'states,' 'processes,' 'functions' and the like: anybody who believes that these things 'exist' (and can enter into causal relations with other things) cannot claim to be a materialist, because materialists, by definition, believe that that the only things that exist are physical bodies. And although obviously 'states' and 'processes' et cetera can *involve* physical bodies, a state/process/function, in itself, is not a physical body.

The reification fallacy is also endemic in virtually any talk about 'philosophy of mind.' For instance, Smithies (2011: 17) talks about whether "experience nonconceptually represents objects as categorical," whereas strictly speaking 'experience' (which is not a causal entity) cannot 'represent' anything, whether that representing is done 'conceptually' or 'nonconceptually.' As I will discuss further in Sect. 4.2.1.3, Boag even claims that "Freud's systemic view of unconscious mentality … reifies the cognitive relation" (Boag 2015: 241)—without apparently noticing that the very use of that phrase 'the cognitive relation' constitutes a prime example of vicious reification (which means that it is highly implausible for him to accuse *Freud* of 'reifying' anything).

My claim that talking about 'the mind' is an example of vicious reification really amounts to no more than the rejection of dualism. It is the same as saying that there is no such 'thing' as 'the mind,' because only (physical) material objects exist in the world. And as I mentioned [Sect. 2.2], endorsement of this claim is constitutive of the materialist position. Papineau claims that the fact we continue to talk about 'the mind' and 'the body' (which I have just said is the same as saying that we continue to *reify* 'mind') reveals that "we are all in the grip of an intuition of dualism"—and that "some of those who profess materialism at a theoretical level … are closet dualists" (2011: 11). I have claimed that anybody who talks about 'the mind' as if it were a 'thing' (comparable to material things) is committing the fallacy of misplaced concreteness; and that anybody who commits the fallacy of misplaced concreteness (or 'reification) with regard to 'the mind' must really be a dualist, whatever s/he might say to the contrary.

Accordingly, if my argument in this section has been successful, not only should the distinction between 'the mind' and 'the body' be viewed as spurious: the distinctions between property/substance dualism [Sect. 2.2.1.3], between reductive/non-reductive materialism [Sects. 2.2.2.1 and 2.2.2.2], and even the distinction between (property-/substance-) *dualism* and (non-/reductive) *materialism* should also be so viewed. In that case, claiming that there is no such thing as a (purportedly) materialist account of the explanatory gap is the same as claiming that there is no such thing as an overtly dualist account—which I have already claimed [Sect. 2.2.1.4]. And if there is no such thing either as a coherent dualist account, or as a coherent materialist account of the explanatory gap, and if these two positions are mutually exclusive and jointly exhaustive, then it follows that it is not possible to produce a coherent metaphysical account of the explanatory gap *at all*.

On the other hand, true physicalists, as we will see in Chap. 5, know perfectly well that there is no such thing as 'the mind' (in the sense of something which is different from 'the body'), which is why they typically do not use such empty terms as 'mental' or 'physical' in their accounts. On a truly physicalist view, there is no such 'thing' as 'the mind'—therefore there should not be even the *illusion* of an 'explanatory gap' between 'the mind' and 'the body.'

2.3 The Explanatory Gap in Twentieth-Century Philosophy

We saw [Sect. 2.2.1] that Descartes was the one who was largely responsible for our idea that there is some kind of deep, metaphysical 'explanatory gap' between physical bodies and thinking things. I claimed that anybody who considered 'the mind' to differ in principle from 'the body' must be a dualist (of either the overt, or the 'closet,' variety); I said that, because dualism was not an acceptable view, we therefore had no satisfactory metaphysical account of the explanatory gap.

In this section, I will discuss four specific examples of work which appear to have had a strong influence on a number of other writers in this area [Sects. 2.3.1, 2.3.2, 2.3.3, and 2.3.4]. However, for the purposes of this chapter—namely, investigating the explanatory gap—it is not necessary that these particular four books or their authors should have been more important or influential than any other four. In fact, if I am correct in saying that the overwhelming problem in philosophy of mind is that the (vicious) reification of 'mind' amounts to the endorsement of dualism [Sect. 2.2.3], I would expect to be equally able to apply my analysis of this problem to any other philosophical position.

In each case, I will consider, firstly, the specific aspect of the explanatory gap problem that concerns the writer in question ('The problem addressed'); secondly, the solution to that problem provided by that writer ('[Philosopher's] Solution'); finally, I will provide 'My assessment' of (1) whether that writer has identified an important or interesting problem to address, (2) whether that writer has succeeded in addressing satisfactorily the problem s/he has identified, and (3) whether that writer's proposed solution to that problem is any help with regard to making sense of the explanatory gap.

I will *not* attempt to provide a comprehensive assessment of each writer's views. Insofar as these writers may not have the same concern with the explanatory gap that I have, my complaints do not necessarily even function as criticisms of each writer's actual views—all I want to do here is to provide some specific examples of influential philosophers who have struggled to articulate a coherent position on the explanatory gap without running into the exact same difficulties which, I have argued [Sect. 2.2], render it impossible to produce a coherent metaphysical account of the gap. I am not attempting to assess the overall position of any of these writers, but rather to use very specific and limited claims the authors have made, for the purposes of demonstrating that it is not possible to produce a satisfactory account of 'the explanatory gap' within either a dualist or a materialist framework.

I hesitate to say that my concern is a purely linguistic one, but what I can say is that the claim I am making here is relatively superficial: I am concerned that these writers have not managed to come up with a coherent way of speaking (about thinking things and physical bodies) that does not exhibit some minimal level of closet dualism—and that this is because they continue to *reify* 'mind.'

2.3.1 Gilbert Ryle (1949): *The Concept of Mind*

According to Tanney, one of Ryle's most significant achievements in this work was that he managed to "put the final nail in the coffin of Cartesian dualism" (1949: ix). However, I will suggest that Ryle is no more able than the other authors to help us understand the apparent explanatory gap between physical bodies and thinking things. Ironically, in spite of the fact that many people agree with Tanney that Ryle's work struck a devastating blow against (Cartesian) dualism, I will suggest that Ryle fails to advance our understanding of the explanatory gap precisely because he himself is unable to articulate a position which is not couched in closet dualist terms.

2.3.1.1 The Problem Addressed by Ryle: The Official Doctrine

The Concept of Mind begins with Ryle's description of what he calls 'the official doctrine,' which is prevalent among both laypersons and scholars. The official doctrine says that people have both 'minds' and 'bodies'; it says that, while "[h]uman bodies are in space and are subject to the mechanical laws which govern all other bodies in space" (1949: 1–2), 'minds' are neither extended in space, nor subject to any of the laws which apply to physical bodies. In fact, 'minds' not only have no causal interactions with physical reality, but do not even interact with other minds (for example, Ryle 1949: 3). The official doctrine takes physical or bodily processes to be 'external' and public, while mental processes are 'internal' and private.

> This antithesis of outer and inner is of course meant to be construed as a metaphor, since minds, not being in space, could not be described as being spatially inside anything else, or as having things going on spatially inside themselves. But relapses from this good intention are common and theorists

> are often found speculating how stimuli, the physical sources of which are yards or miles outside a person's skin, can general mental responses inside his skull, or how decisions framed inside his cranium can set going movements of his extremities. (Ryle 1949: 2)

But however 'official' or familiar it may be, the official doctrine is, according to Ryle, "entirely false, and false not in detail but in principle" (1949: 6).

2.3.1.1.1 Category Errors

The official doctrine embodies a mistake of the type referred to by Ryle as a category mistake. Ryle "never explicitly defines the notion of a 'category'" (Schnaitter 1985: 146), instead using examples to make his meaning clear.

One of his examples is as follows:

> A foreigner visiting Oxford or Cambridge for the first time is shown a number of colleges, libraries, playing fields, museums, scientific departments and administrative offices. He then asks 'But where is the University? I have seen where the members of the Colleges live, where the Registrar works, where the scientists experiment and the rest. But I have not yet seen the University in which reside and work the members of your University.' … His mistake lay in his innocent assumption that it was correct to speak of Christ Church, the Bodleian Library, the Ashmolean Museum *and* the University, to speak, that is, as if 'the University' stood for an extra member of the class of which these other units are members. He was mistakenly allocating the University to the same category as that to which the other institutions belong. (Ryle 1949: 6)

Ryle suggests that when people talk about thoughts, feelings, desires, ambitions, et cetera, *and* 'the mind' (as if 'the mind' stood for an extra member of the class of which these other units are members) they are making exactly the same kind of mistake.

2.3.1.1.2 The Dualist Intuition

In Ryle's view, the origin of the category mistake, in the case of the mind and the body, can be traced back to Descartes' *Meditations*. Descartes' problem was that he was both a religious man and "a man of scientific genius" (1949: 8): if he wanted to maintain that persons were created by God, and that they were animated by souls, he would have to give up on his scientific principles; but if he wanted to hold onto his belief in the scientific method, he would be forced to give up his belief in the existence of 'the divine soul.' This is why, according to Ryle, Descartes and subsequent philosophers "naturally but erroneously" (ibid.) decided that 'mental' processes must be different in principle from (physical) mechanical processes—or at least, that at least some of the mechanical movements of human bodies must be produced by non-mechanical causes.

But Descartes was unable to spell out exactly how this could work:

> Still unwittingly adhering to the grammar of mechanics, he tried to avert disaster by describing minds in what was merely an obverse vocabulary … they are not in space, they are not motions, they are not modifications of matter, they are not accessible to public observation. Minds are not bits of clockwork, they are just bits of not-clockwork. (Ryle 1949: 9)

Descartes and his contemporaries were not able to elaborate upon the ways in minds were, but at the same time, were not, like machines—for the very good reason, says Ryle, that they were thinking of 'the mind' and 'the body' as if they were members of a single category, when actually they are members of different categories. Just as, for example, it makes no sense to say that someone bought *either* 'a left-hand glove plus a right-hand glove,' *or* 'a pair of gloves,' it makes no sense to say that *either* a mental process occurred, *or* a physical process occurred (Ryle 1949: 12). Saying that a mental process occurred is just not the same kind of thing as saying that a physical process occurred, and that is why it is inappropriate to conjoin or disjoin the two statements.

2.3.1.2 Ryle's Solution: Philosophical Behaviorism

Ryle claims that if the "mental predicates" (1949: 39) we commonly use to describe people and what they do really did refer to mysterious 'mental' phenomena (which by definition we could neither observe nor otherwise investigate), then such words would be no better than meaningless. The fact that they are not meaningless—that, on the contrary, we rely on these terms as an essential part of our day-to-day lives—shows us that they cannot really be referring to mysterious (private, occult) 'mental' phenomena at all. When we use mental predicates, Ryle claims, "we are not making untestable inferences to any ghostly processes occurring in streams of consciousness which we are debarred from visiting: we are describing the ways in which these people conduct parts of their predominantly public behaviour" (ibid.). This is why many people have described Ryle's theory *philosophical* (or *analytical*) *behaviorism.*

2.3.1.3 My Assessment

2.3.1.3.1 Category Errors/The Dualist Intuition

Ryle has recognized that 'the mind' is not an independent causal entity (independent of the thoughts, feelings, beliefs, projects, et cetera that comprise it), just as 'a pair of gloves' is not an independent entity which is either independent of, or in some sense different from, the left-hand and right-hand gloves that comprise it. His position regarding The Category Mistake inherent in the Official Doctrine is exactly right: when we say that thoughts/feelings et cetera comprise 'the mind,' we are making a conceptual (category) error. Speaking as if 'thoughts/feelings/et cetera' and 'the mind' are different things is just as bad as speaking as if 'pairs of gloves' are different from 'left-hand-plus-right-hand gloves.'

2.3.1.3.2 *Philosophical Behaviorism*

Ryle's concern is not whether 'the mind' is different in principle from 'the body' (or whether dualism or materialism provides the better account of what-it-is-like/experience), but whether it makes sense to think that there is even such a thing as 'the mind' (or 'mental conduct') in the first place. Therefore, worrying about whether Ryle's work did or did not finally 'put the final nail in the coffin' of dualism may be to miss the most important point, because what Ryle actually does is draw our attention to the severe problems inherent in any account which construes 'mental conduct' as any kind of 'inner' process—regardless of whether these inner processes are "*irreducibly mental or at bottom physical*" (Tanney 2015, Sect. 6.2, my emphasis).

Ryle claims that everything we usually refer to using 'mental concept terms' can be satisfactorily accounted for by referring only to behaviors, and dispositions to behave. But this is problematic. I would suggest that the majority of our 'behaviors'—even when 'behaviors' are understood to include dispositions to have thoughts and feelings—are unique to particular situations: specific, once-off reactions. By definition, 'dispositions' (defined as 'tendencies to such-and-such') are only comprehensible in the context of a certain amount of repetition—which means that any kind of mental conduct which is non-recurring is not even potentially amenable to a dispositional explanation.

2.3.1.3.3 *Ryle's Reification*

Although Ryle himself does not use the term 'reification,' it seems to me that at least some of the examples of what he calls category errors are also instances of what I am calling reification. For example, when somebody refers to *a pair* of gloves, s/he is referring to a thing (a pair) which does not exist in (physical) reality, but only in language. What exists in reality is two objects, which we refer to as gloves. Describing the two objects as a *pair* of anything is purely a conceptual act: it is like referring to a certain group of meteorological phenomena as a hurricane. Strictly speaking, and if she was unable to fall back on accepted linguistic conventions, that person would have to say that there were two specific objects (a left-hand glove and a right-hand glove) which she had misplaced (and not 'a pair of gloves')—just as, strictly speaking, it is certain amounts of wind and/or water, being moved by certain kinds of force, and not 'hurricanes,' that can be said to cause damage to trees. 'Hurricanes,' like 'pairs,' are things which exist only in language, and not in the (physical) world. So in this particular case, Ryle's category error and my '(innocent) reification' seem to coincide.

However, although Ryle noticed the problem of vicious reification (which is the form taken by some of his 'category errors'), he nevertheless commits quite a bit of it himself. For instance, referring to "the consciousness which irradiates [mental states and processes]" (Ryle 1949: 3) is extreme vicious reification, because 'consciousness' is not a 'thing' which can 'irradiate' anything else. That would be like saying that 'pairhood' 'irradiates' the left-hand and right-hand gloves. Even more alarmingly, Ryle says (in a discussion about what it means to 'theorise') that "in theorising the soul is ... preparing itself to talk or write didactically" (1949: 262)! And as we know [Sect. 2.2.1], the only people who talk about 'the soul' as a causal entity are extreme Cartesian dualists. Such is the strength of 'the dualist intuition' [Sect. 2.3.1.1] that Ryle turns out both to be claiming that he is strongly opposed to dualism, and implicitly committing himself (by referring to 'consciousness' and 'the soul') to a strong form of dualism.

2.3.1.4 Gilbert Ryle and the Explanatory Gap

The problem was that there appeared to be an 'explanatory gap' between 'the mind' and 'the body.' Now Ryle has explained why there appears to be this gap (we are making a category error when we referred to 'the mind' and 'the body' as if they were different members of the same category); and he has given us a new way of speaking, so that this no longer appears to be the case (we are to talk about 'behavior' and 'dispositions to behave,' rather than about mysterious or occult 'mental' conduct). However, if we do not *wish* to speak about behavior, and dispositions to behave—if we still want to understand why there is *something-it-is-like* to behave in certain ways—none of this is any help at all. Rather than answering this question, Ryle has simply told us that it is not a question we should be asking in the first place [compare: Hill 2010a: 6]. If what we were looking for was a conceptual understanding of the explanatory gap, Ryle's account has not provided us with such an understanding.

2.3.2 Patricia Churchland (1986): *Neurophilosophy*

Robert Paustian, in his Editorial Review (*Library Journal*, 1986) had this to say about Churchland's *Neurophilosophy: Toward a Unified Science of the Mind-Brain*:

> This profound, exhaustive, and well-written scholarly work is an excellent introduction to neuroscience from a philosopher's point of view. ... In demonstrating the relevancy of neuroscience to philosophy, Churchland ... argues that ... detailed knowledge of the organization and structure of nervous systems is necessary for the evolution of an adequate theory of the mind/brain.

Neurophilosophy was among the first attempts to integrate philosophy and neuroscience, and it had a strong impact on much of the thinking in late twentieth and early twenty-first century philosophy of mind. However, I am claiming that—largely because it seems to exhibit closet dualism—Churchland's 1986 work does not bring us any closer to making sense of the explanatory gap.

2.3.2.1 The Problem Addressed by Churchland: Acceptable Theories

Churchland sets out "to paint in broad strokes the outline of a very general framework suited to the development of a unified [causal] theory of the mind-brain" (1986: 3). She takes it for granted that "nervous systems are information-processing machines" (1986: 36); therefore, it follows that in order to outline this general framework, what is urgently needed is to develop "[n]ew theories about the nature of *information processing and about the nature of information-bearing structures*" (1986: 399, my emphasis).

2.3.2.2 Churchland's Solution: Tensor Network Theory

According to Churchland, the 1977 study by Pellionisz, Llinás and Perkel "illustrates some important aspects of the problem of theory in neuroscience," and might therefore be just the kind of "general framework for understanding the computational architecture of nervous systems" (Churchland 1986: 411–412) that Churchland is looking for.

The cerebellum (meaning 'little brain') is quite an old brain structure, in terms of its production during the evolution of our species; and is mostly important for regulating things to do with balance, posture, and voluntary movement. According to Churchland, "almost everything one would want to know about the micro-organization of the cerebellum is known" (1986: 412), which is what makes investigation of the cerebellum the ideal starting place for anybody who wants to develop a theory about the computational architecture of nervous systems. Pellionisz, Llinás and Perkel created a simulated cerebellum by programming a computer to 'grow' a large number of (imaginary) cells, and create connections between the cells, to approximate the real behavior of cells in the cerebellum. The main finding of this study was that "the connectivity of arrays of neurons is crucial to explaining how a given input yields a given output" (Churchland 1986: 416). The authors realized that they could represent this input-output relation in terms of a 'tensor,' or "generalized mathematical function for transforming a vector into another vector, no matter what the frames of reference involved" (1986: 417). This is what Churchland refers to as 'the tensor network hypothesis': the hypothesis which says that "a neuronal network implements its general function as a connectivity matrix to transform input vectors into output vectors" (1986: 432).

Churchland suggests that the tensor network approach can be useful when applied to 'problems' such as sensorimotor control (that is, "controlling behavior on the basis of sensory information," Churchland 1986: 450), and 'gaze control' (that is, the different operations our brains perform during visual experience to provide us with the sense of looking at a stationary object, even as we ourselves are moving; or the sense of looking at a single object in motion, rather than a progression of stationary objects). Calculating exactly how much each muscle in the eye, neck, shoulder et cetera needs to contract in order to accomplish this is, on Churchland's view, just the sort of thing that a three-dimensional 'vector' type of calculation would be good at: transforming "the vestibular vector" into "the ocular vector" (1986: 436).

2.3.2.3 My Assessment

2.3.2.3.1 Nervous Systems vs Computers

I can fully endorse Churchland's suggestion that we are in need of a new paradigm for understanding 'the mind-brain.' But then she apparently just takes it for granted that such a paradigm must, as a matter of logic, comprise a theory about something to do with *information processing* [Sect. 2.3.2.1]—which is problematic. When the first computers were produced—or the first personal computers, or the first incredibly powerful computers suitable for calculating advanced mathematical formulae—everybody was so awed by the vista of possibilities that seemed to be opening up that it was not always easy to remember that the computer is really nothing more than a useful tool (Searle 1997: 190). It is undeniably a much more powerful and useful tool than, for example, the pocket calculator; but what it is not is a tool that is different *in principle* from the pocket calculator. Churchland admits that "[o]nly in a *very* abstract sense is the brain like a computer" (1986: 408), but we have already seen [Sect. 2.3.2.1] that she takes it for granted that 'nervous systems are information-processing machines': it certainly seems that using the computing metaphor quite literally is exactly what she herself is doing, in spite of cautioning us to remember that brains and computers are similar only in a very abstract sense. In fact, it is reasonable to suggest that nervous systems *are not* really all that similar to computers. Perhaps the most devastating problem with the computational framework is "how profoundly *antibiological* it is" (Searle 1997: 190): it seems likely that many of the functions that brains perform "simply cannot be computed in any credible manner" (Panksepp 2003: 8).

There is at least one way in which computers invariably perform better than organisms. When a neuron is stimulated, it takes a certain (short) period of time for that neuron to revert to its original, pre-stimulated condition; during this period of 'fatigue,' that neuron can "not immediately accept stimulation by another pulse and respond in the standard way" (Von Neumann 1958: 46). This means that the reaction time of neurons is slower, by a factor of about 10^4 to 10^5, than the reaction times possible in artificial components. In virtually all other respects, however, natural systems perform better than artificial ones. With respect to the linear size (of the cell membrane) needed to produce a specified amount of processing power, "the natural components seem to lead our artifacts by a factor like 10^3." With respect to volume requirements (that is, the space taken up by necessary processing components), "the natural components lead the artificial ones … by factors like 10^8 to 10^9" (Von Neumann 1958: 48). The amount of energy consumed by natural components is greater than the energy consumed by artificial components by a factor of 10^8 to 10^9 (Von Neumann 1958: 49). Finally, "in terms of the number of actions that can be performed in a certain interval, natural systems typically outperform artificial systems by a factor of 10^4" (Von Neumann 1958: 50).

We are not sure exactly how biological systems manage to produce such fast and accurate responses, but there are a number of respects in which organisms effortlessly outperform even the most sophisticated computers. Rather than being similar to computational processes, Panksepp thinks that brain processes are "deeply organic and analog" (2003: 8). Whereas computers function by means of following rigid logical rules, biological organisms function in accordance only with the 'logic' of evolution by natural selection—and the logic of evolution by natural selection is very different from the 'logic' of 'computation.' So when Churchland sets out to discover new theories about the nature of information processing, in the hopes that this may be useful for developing a theory of the mind-brain, she may not be looking in the right place.

2.3.2.3.2 Churchland's Reification

Churchland says that "[n]euroscience may fail to explain phenomena characterized in folk psychology for the same reason Newtonian physics fails to explain what turns the crystal spheres and modern biology fails to explain how the vital spirits are concocted" (1986: 385). How ironic it is that Churchland has made this remark, at this point—because this is the very charge I am trying to lay against her: that she is unable to produce a satisfactory 'answer' to her question, because she is asking the wrong question! She says that "[h]ow to characterize the mathematical relationships between the response profiles of the input and output ensembles is not something that in effect falls out of an array of data, though it may well be inspired by it" (1986: 406)—without, apparently, considering the possibility that *there are no such relationships* between 'response profiles of the input and output ensembles' *because there are no such 'things' as response profiles, or as input/output ensembles, in the first place*!

Unlike Ryle, Churchland may not be guilty of thinking that 'consciousness' *irradiates* 'the mind'; but her reification of information-processing-related concepts (like response profiles, input/output ensembles et cetera) is just as damaging for the credibility of her position as the reification of mind/soul/thinking thing is for Cartesianism. On her account, we have, on the one hand, things like the (physical) brain; and on the other, things like 'information' and 'response profiles,' which are different in kind from physical bodies. Because she understands these two classes of things to be significantly different (in principle), I suggest that Churchland's Tensor Network Theory can be seen as an instance of "a kind of last gasp of … the residual dualism" (Searle 1997: 192) which is typical of computational accounts. Clearly, unlike true physicalists, she believes that at least some non-physical things can genuinely be said to 'exist.'

2.3.2.4 Patricia Churchland and the Explanatory Gap

It not plausible to try to describe biological organisms in computational terms [Sect. 2.3.2.3]. In fact, attempting to characterize the mathematical relationships between the response profiles of input/output ensembles (which is what the Tensor Network Theory professes to do) sounds very much like attempting to characterize the relationships between the library-and-playing-field profiles of university ensembles: a category error (and/or vicious reification). Moreover, even if it were plausible to use a Tensor Network-type account on human persons and their what-it-is-like/experience, this would not get us any closer to closing the explanatory gap.

> In itself, functionalism is simply a theory of the meaning of the concept of mind.
>
> However, functionalism is not, and cannot be a theory of the *basis* of the human mind. Such a theory must answer the question: 'What are the empirical conditions for the possibility of those input-internal-change-and-output patterns called the human mind?' (Wiredu 2002: 62)

Because Churchland does not answer the question about empirical conditions, her view cannot tell us anything about how to close the explanatory gap.

2.3.3 David Chalmers (1995): *Facing Up to the Problem of Consciousness*

Chalmers is well-known for dividing 'the problem of consciousness' into 'the easy problems' and 'the hard problem' (which I will discuss further in Chap. 3). He is of the opinion that our understanding of 'the easy problems' of consciousness is relatively comprehensive, and getting better all the time. The only thing really lacking in this regard is a little more working out of the 'details.' On the other hand, he thinks that "there is nothing that is harder to explain" (1995: 1) than 'the hard problem' of consciousness. But Chalmers is a self-confessed dualist—and, as I have stipulated [Sect. 2.2.1.4], I will not consider dualism to be a satisfactory framework within which to try and account for the explanatory gap.

2.3.3.1 The Problem Addressed by Chalmers: The Hard Problem Persists

The easy problems of consciousness are problems which "seem directly susceptible to the standard methods of cognitive science" (Chalmers 1995: 2). For example, 'reportability' can be understood as constituting those neural or computational mechanisms that perform the verbal report function. Specifying these neural/computational mechanism/s is all that 'explaining' this function consists of. Other easy problems might be things like "the ability to discriminate, categorize, and react to environmental stimuli," "the integration of information by a cognitive system," "the ability of a system to access its own internal states," and "the deliberate control of behavior" (ibid.).

Whereas there are many 'easy problem' of consciousness, there is really only one 'hard problem'—and that is the problem of experience. Although Chalmers thinks that we can quite easily 'explain' all functions like reportability in terms of neural or computational mechanisms, what we are not readily able to explain is why there is something-it-is-like to perform those easy-problem functions. This problem "persists even when the performance of all the relevant functions is explained" (1995: 4); it is a "nontrivial further question" (1995: 6). But it is a nontrivial further question which cannot even potentially be answered by doing empirical investigation; in fact, in Chalmers' view, we are yet to come up with even a single really good way of addressing this problem.

> Quite a bit of work on consciousness has appeared in the last few years [but] most of this work leaves the hardest problems about consciousness untouched. (Chalmers 1995: ix)

2.3.3.2 Chalmers' Solution: Taking Experience as Fundamental

In fact, according to Chalmers, *there is no* really good way of explaining why the performance of certain functions is accompanied by what-it-is-like/experience; so rather than trying to explaining why there is something-it-is-like to perform the (easy-problem) functions, he thinks we should just accept this as a brute fact. He says that, "[a]lthough a remarkable number of phenomena have turned out to be explicable wholly in terms of entities simpler than themselves, this is not universal" (1995: 13). Even in a community which approves of breaking down complex concepts into ever-more-fundamental components, at some point, it always becomes impossible to break down any further whatever phenomenon one is considering. Mass and space-time are usually taken to be 'fundamental' in this sense: nobody even attempts to break them down into anything more fundamental than they already are.

In just this way, suggests Chalmers, "a theory of consciousness should take experience as fundamental" (1995: 13–14)—as the most ontologically-basic unit of explanation, which cannot be broken down any further. He thinks it is 'obvious' that our basic (materialist) ontology needs to be supplemented in some way, since nothing we know about physical principles shows why any of them should 'give rise to' 'consciousness.' And if our ontology needs to be supplemented, why not supplement it in the most obvious and straightforward way—by adding an extra 'fundamental category,' namely, 'experience'? Just as nineteenth century scholars determined that electromagnetic processes "could not be explained in terms of the wholly mechanical processes that previous theories appealed to" (1995: 14), so, Chalmers thinks, should twentieth century philosophers just accept that 'experience' cannot be explained in terms of any kind of other, simpler components.

2.3.3.2.1 Information Processing

Chalmers then argues that if we take experience as fundamental, "a naturalistic account of consciousness can be given" (1995: 1). In his view, such an account should be based on (1) "principles of structural coherence"; (2) "principles of organizational invariance"; and (3) "a double-aspect view of information" (1995: 1). I will not examine these principles individually; suffice it to say that, on Chalmers' view, the important thing is establishing "a fundamental link between information and experience."

Once this link has been established, one can safely say that "phenomenal properties are the internal aspect of information" (1995: 24).

2.3.3.3 My Assessment

2.3.3.3.1 Hard/Easy Problems

Chalmers proposed that the easy-problem phenomena were all straightforwardly susceptible to explanation in terms of computational or neural mechanism. I agree that it is useful to distinguish between the two types of problems, and I will use the terms 'hard problem' and 'easy problems' throughout; however, it is not strictly true that easy-problem phenomena can be 'explained' in terms of computational or neural mechanisms. Strictly speaking, all that neuroscience can be used to do in these cases is establish *correlations*. When research participants view objects in the left-hand part of their visual field, for example, we know that we will reliably observe certain types of activity in a certain part of the primary visual cortex of the brain in the contralateral hemisphere. But this is not quite the same as *explaining* that visual experience in those neural terms. It is not an 'explanation,' but a brute fact: when *a* happens, *b* (also) happens. We know *what happens in the brain* when we have certain sensory perceptions, but we don't know *why* those things in the brain are experienced in this way. Discovering a *correlation* between 'consciousness' and a certain type of neural activation is not enough: additionally, we would need to ask ourselves "what would it signify if a specific pattern of neural activation was empirically confirmed to correlate with consciousness?" (Montemayor and Haladjian 2015: 73).

Therefore, although we might be able to see the value of separating 'the easy problems' from 'the hard problems,' in the name of improving the conceptual clarity of our discussions, we should probably not agree with Chalmers that *any* of these ('easy' or 'hard') problems are 'susceptible to *explanation* in terms of computational or neural mechanisms.' Even the easy problems are harder than Chalmers suspected.

2.3.3.3.2 *Taking Experience as Fundamental*

Having decided that the way forward is to take 'experience' as a fundamental category, Chalmers proceeds to set out his vision of what a non-reductive theory of experience might look like. He says that this theory will specify basic principles describing the way in which 'experience' relates to 'physical features of the world.' He says that "[t]hese *psychophysical* principles will not interfere with physical laws, as it seems that physical laws already form a closed system" (1995: 14). He says that those psychophysical principles, having refrained from 'interfering' with those physical laws, will, rather, 'supplement' physical theory. But this is problematic. Either mental phenomena are able to interact in some way with physical phenomena, or they are not. If the 'mental' phenomena *do* interact with physical phenomena, then the principle of causal closure is violated, and the theory must be rejected by anybody who takes dualism to be problematic. If the 'mental' phenomena *do not* interact with physical phenomena, then there does not appear to be any way in which 'psychophysical principles' can 'supplement' physical theory, or in fact be anything other than causally irrelevant and epiphenomenal.

2.3.3.3.3 *Information Processing*

Chalmers not only thinks that 'experience' is a 'thing' which 'exists,' and therefore can be investigated (that is, he is guilty of vicious reification): he also thinks, in particular, that a naturalistic account of 'consciousness' must be based on a certain view of 'information.' However, as I have already explained at some length [Sect. 2.3.2.3], it is not plausible to suggest that the way nervous systems function is in any way analogous to the way that computers function (or 'process' 'information'). Moreover, 'information' does not name a real 'thing' (that is, using 'information' in a causal sense is an example of vicious reification): something can only ever be described as 'information' insofar as it plays a specific role in the conceptual understanding of a certain kind of linguistic agent. Therefore, (seeing that there is no such 'thing' as 'information'), there can be no such thing as a 'double-aspect view of information,' or such a thing as 'the internal aspect of information' [Sect. 2.3.3.2.1]—which leaves Chalmers' account very light on legitimate content.

2.3.3.3.4 Chalmers' Reification

According to Chalmers, cognitive science and neuroscience are unsuitable tools for explaining conscious experience; however, "[i]t is not too soon to begin work on a theory. We are already in a position to understand certain key facts about *the relationship between physical processes and experience, and about the regularities that connect them*" (1995: 16, my emphasis). But there is nothing innocent about this reification. Talking about 'the relationship between physical processes and experience, and the regularities that connect them' is like talking about 'the relationship between meteorological processes and hurricanes, and the regularities that connect them' (or 'the relationship between the colleges, libraries, playing fields et cetera and universities, and the regularities that connect them'; or 'the relationship between right-hand and left-hand gloves and the pairhood which irradiates them'). There are no actual relationships between any of these things: the only 'relationships' between them are linguistic in nature, resulting from the way we, as a linguistic community, refer to these things in our language [Sect. 2.2.3.1]. Chalmers never tries to deny that his position is a dualist one, but he claims that his is "an innocent version of dualism, entirely compatible with the scientific view of the world" (even going so far as to say that *naturalistic dualism* might be a good choice of name for this view).

2.3.3.4 David Chalmers and the Explanatory Gap

According to Chalmers, "[n]othing in this approach contradicts anything in physical theory" (1995: 15)—but I can see precious little difference between saying 'The body exists, in a normal, physical sense, whereas the mind exists in some different/non-material sense' (which is what Cartesian dualists say), and 'The body exists in a normal, physical sense, whereas *experience* exists in some different/non-material/fundamental sense' (which is what Chalmers says). These are not plausible claims. Perhaps we could say that trying to find out why and how experience arises from a physical basis is like trying to find out why and how the university arises from the colleges, the library and the playing fields: a category error (and/or vicious reification). Therefore, Chalmers' work does not help with making sense of the explanatory gap.

2.3.4 John Searle (1997): The Mystery of Consciousness

In *The Mystery of Consciousness*, Searle turns his "sharp eye for confusion and contradiction" (www.amazon.com/Mystery-Consciousness-John-R-Searle/dp/0940322064) towards possible ways of conducting future consciousness research within a naturalistic framework. Although Searle refers to his view as 'biological naturalism,' and evidently intends to frame his theory in 'naturalistic' terms, in my opinion, this is not enough to insulate him against my charge of closet dualism; therefore, unsurprisingly, Searle's view turns out to be not much help to anybody who is interested in making sense of the explanatory gap.

2.3.4.1 The Problem Addressed by Searle: Obsolete Categories

According to Searle, "the greatest single philosophical obstacle to getting a satisfactory account of consciousness is our continued acceptance of a set of obsolete categories" (Searle 1997: xi-xii)—he is thinking particularly of the categories 'mental/physical,' 'dualism/monism,' and 'materialism/idealism.' We not only assume that these terms really do carve nature at its joints, but also that they are the most correct—perhaps, indeed, *the only*—terms in which to pose our questions. But Searle thinks that our reliance on these categories "creates much of our problem and does not point the way to a solution" (1997: 210–211).

For instance: it appears, to "[m]any mainstream philosophers" (Searle 1997: 194), that if we endorse materialism (which is to say, if we consider dualism to be problematic), we cannot simultaneously believe in *anything* which is non-physical—and this includes not only supernatural (in the sense of *divine*) forces, but also all kinds of *mental conduct* terms. On the other hand, if we want to hold onto our belief in ordinary 'mental' concepts, it seems we can't simultaneously be materialists. Seeing that the acceptance of either dualism or materialism forces us to give up a central part of our common-sense worldview, Searle thinks that both positions must be fatally flawed.

He thinks that "if you have a map on which there are only two mutually exclusive territories, the 'mental' and the 'physical,' you have a hopeless map and you will never find your way about" (1997: 195). In much the same way Ryle pointed out that it made no sense to ask whether something was *either* an actual pair of gloves *or* merely a left-hand glove plus a right-hand glove, so (suggests Searle) it makes no sense to think that everything there is must be *either* mental *or* physical. On the contrary, the world "contains all kinds of things—money, interest rates, reasons for voting for or against the Democratic candidate, laws of logic, points scored in football games—that are not in any obvious sense either mental or physical" (1997: 88).

2.3.4.2 Searle's Solution: Non-event Causation

On Searle's view, the way we understand 'cause' and 'effect' could also be an example of obsolete (and falsely dichotomous) categories. The materialist is supposed to believe that everything that exists is physical—therefore, this materialist does not have the option of saying that a (conventional) causal relation exists between 'brain processes' and 'mental processes' (seeing that, on a materialist view, 'mental processes' do not exist in the first place). But Searle points out that the kind of cause-and-effect relation that we typically think of is not the only possible kind of causation which can be accommodated within a naturalist framework. We think of causes in the sense of *events*, in the sense in which we can say 'the earthquake in the Pacific ocean caused the tsunami in the Philippines.' This is what Searle refers to as 'event causation.' One event occurs, and it has tangible physical effects, and these effects in turn cause the next event.

However, there is a different kind of causal relation that is also familiar. Searle tells us to "think of the causal explanation of the fact that the table exerts pressure on the rug … Or … the solidity of the table" (1997: 7). Gravity 'explains' the pressure of the table on the rug, and the behavior of the molecules in the table 'explains' its solidity; these explanations can be referred to as *causal*, but these causes are not *events* (in the sense that the earthquake that caused the tsunami is an event). These examples of non-event causation "give us appropriate models for understanding the relation between my present state of consciousness and the underlying neurobiological processes that cause it" (1997: 7–8).

Searle thinks that there must be "a discoverable causal relation" (1997: 195)—even if the causation in question is of the 'non-event' variety—between 'consciousness' and certain brain processes; and that, therefore, our first step must be to find a correlation between the two. Our second step must be establishing the direction of causality: finding out whether initiating or terminating those brain processes can be shown to initiate or terminate the corresponding conscious experiences in a reliable way. To this end, he suggests that "a promising line of attack" (1997: 199) would be to focus on clinical cases where conscious experience is in some way disturbed or disrupted. For example: *blindsight* refers to the condition (which I will be discussing at length in Sect. 3.2.1) in which damage to a specific part of the visual processing region of the brain results in patients' being able to *respond* to visual stimuli in certain limited respects, without having any awareness of being able to *see* the stimuli. Searle suggests that blindsight might be a paradigm example of such a case of 'disrupted' conscious experience: a case where 'consciousness' itself is missing from that visual experience.

'Consciousness' is "a natural biological phenomenon" (Searle 1997: xiv); but what makes 'consciousness' different from other natural biological phenomena is that is *caused* by ('lower level') processes of the brain, even as it is *a feature of* the brain (at the 'higher' level). Moreover, "[c]onsciousness is not reducible in the way that other biological properties typically are, because it has a first-person ontology … [C]onsciousness only exists when it is experienced as such" (Searle 1997: 213).

2.3.4.3 My Assessment

2.3.4.3.1 Obsolete Categories/Dualism vs Materialism

Searle, like Ryle, has recognized that there is some kind of conceptual mistake going on when we talk about 'the mind' and 'the body' as if they belonged to different categories. On Searle's view, the problem with the way we talk about 'conscious states' and 'neurobiological states' is that *we are using obsolete (and/or falsely dichotomous) categories*, which seems to be virtually identical to what Ryle meant when he said *we are making a category mistake* (and what I meant when I said how commonly people fall into the trap of committing vicious reification). However, although Searle claims that it makes no sense to say that everything there is must be either mental or physical, he does not, for instance, claim that *everything there is is physical*, or that *there is no such thing as* something which is not physical (that is, which is mental). He is not denying that it is meaningful to distinguish 'mental' from 'physical' things: just denying that the dichotomy of 'mental' and 'physical' exhausts all the options. But the claim 'everything there is is physical' (or that there is no such thing as something which is not physical) is supposed to be the central commitment of materialism [Sect. 2.2.2]—whereas recognizing that 'mental' and 'physical' things are different in principle is constitutive of dualism [Sect. 2.2.1]. Searle's claim that his position is a 'naturalistic' one is therefore spurious.[6]

2.3.4.3.2 Non-event Causation/Blindsight

Searle said that we need to discover a causal relationship between brain processes and consciousness; he suggested that 'a promising line of attack' for identifying such a causal link would be the investigation of "processes in the brain that are psychologically real but which have no conscious manifestation" (1997: 199); like, for example, blindsight [Sect. 2.3.4.2]. However, the fact that Searle thinks that 'psychological processes' and 'processes in the brain' are distinct enough to enjoy a causal relation clearly demonstrates his (closet) dualism.

Moreover, although Searle's examples (about marks on rugs caused by the pressure exerted by the legs of tables) illustrate very clearly his idea of non-event causation, it is not immediately obvious how this kind of example could be applied to the question of the 'mind' and the 'brain.' In his example, although he didn't make this explicit, there are really three essential constituents involved in the production of the indentations on the rug: the table, the rug, *and the force of gravity*. On its own, the table would not 'cause' the indentations on the rug, either in an 'event causation' sense or in a 'non-event causation' sense: it would only do so in the presence of the right kind of metaphysical force. But what kind of analogous metaphysical force could we possibly identify by means of which 'the brain' could exert any kind of (non-/event) causal influence on 'consciousness'? It seems to me that the short answer to that is 'There is no such force'—*for anybody who is not a dualist!*

2.3.4.3.3 Searle's Reification

Therefore, although Searle maintains that his position is naturalistic (and although this claim is certainly more plausible than Chalmers' claim that his *dualist* position is 'naturalistic'), we should probably still reject this claim. Searle rejected the false dichotomy of 'mental' and 'physical,' and suggested that we should devise a theory of 'consciousness' as a biological phenomenon—both of which things are commendable. Unfortunately, the concept of 'event causation' was not able to do the work Searle wanted it to with regard to dissolving this dichotomy. He wanted to spell out a way in which 'the mind' could have a legitimate causal influence on 'the body'—but in order to produce any kind of causal relation, of the event-causation or the non-event-causation variety, it is necessary to have at least two discrete entities ('the mind' and 'the brain') engaging in this causal relation. Therefore, insofar as he takes it for granted that there is a difference between 'mental' and 'physical' types of causation, Searle manifests a (closet) dualist mindset, in spite of aiming for a 'naturalistic' account.

2.3.4.4 John Searle and the Explanatory Gap

Saying that 'consciousness is not reducible in the way that other biological properties typically are, because it has a first-person ontology' [Sect. 2.3.4.2] sounds suspiciously similar to saying 'pairhood is not reducible in the way that other glovely properties typically are, because it has a left-and-right ontology.' And saying that 'consciousness is a natural biological phenomenon that … is caused by lower-level micro-processes in the brain and it is a feature of the brain at the higher macro levels' sounds suspiciously like saying that 'universitihood is caused by libraries, while at the same time universitihood is a feature of libraries at the higher macro levels': it is a category error (and/or vicious reification). Therefore, even if Searle's account of non-event causation were plausible, it certainly gets us no closer to understanding the explanatory gap.

2.3.5 The Explanatory Gap in Twentieth-Century Philosophy: Conclusion

Of the four influential twentieth-century philosophers I discussed, the only one who took an overtly dualist position was Chalmers; but the other three authors, who claimed to be physicalists/materialists, continued to talk about 'the mind' (and/or 'consciousness' and/or 'experience' et cetera) and 'the body' (and/or the brain and/or nervous system et cetera)—which seems to be a good indication that, on their views, 'the mind' not only exists, but is also different in principle from 'the body.' This way of speaking, I suggested (following Papineau, see Sect. 2.2.3), is indicative of dualism. And dualism is not a satisfactory framework within which to account for the explanatory gap [Sect. 2.2.1.4].

The severe problems inherent in both dualist and materialist (which is to say, both overtly dualist and 'closet' dualist) frameworks also render invalid many philosophical attempts to make sense of the explanatory gap. None of the four typical accounts I examined seems to contain an answer to the question of why there appears to be this (explanatory) gap between mind and body—let alone how to close this gap. I have argued that the only people who can even articulate such a concept as 'the explanatory gap' must be (at least 'closet') dualists—and I have stipulated that I will not consider dualist views as being plausible.

Because neither dualism nor materialism can make any sense of the explanatory gap—and especially because even supposedly 'materialist' accounts often turn out to be dualist anyway—it is not possible to produce a coherent conceptual/philosophical account of the explanatory gap.

2.4 The Explanatory Gap: Conclusion

In this chapter, I have demonstrated that:

1. Because neither dualism nor materialism can make any sense of 'the explanatory gap'—and because dualism and materialism (in the limited senses in which I am considering them, for the purposes of this book) are mutually exclusive and jointly exhaustive—it is not possible to produce a coherent metaphysical account of the explanatory gap [Sect. 2.2]; and
2. Because neither dualism nor materialism can make any sense of 'the explanatory gap'—and especially because even supposedly 'materialist' accounts very often turn out to be dualist anyway—it is not possible to produce a coherent conceptual/philosophical account of the explanatory gap [Sect. 2.3].

So many philosophers spend so much time arguing with one another about how to close the explanatory gap—without apparently considering that *none of* their attempts can possibly succeed, *because 'the problem of the explanatory gap' is a false problem*! The only people who can talk about 'the mind-body problem' are people who believe that 'the mind' *exists* in the first place, and I have argued that it is not possible to believe this unless one's view is at least to some minimum degree a dualist one. And I have stipulated that, for the purposes of this book, I will not consider any kind of dualist view to be plausible. It is not possible to *account* for 'the explanatory gap' between 'the mind' and the body, because it is not possible, on either a dualist or a materialist view, to articulate how 'mental' phenomena could *exist* within a physical world.

Accordingly, I will now abandon my investigation of 'the problem of the explanatory gap': the rest of this book is devoted to searching for a satisfactory alternative way of understanding how thinking/consciousness is un-mysteriously produced during the course of the functioning of physical structures. By the end of Chap. 6, I hope to have demonstrated convincingly that on what I refer to as a 'biological naturalist' view, there should be no temptation for anybody to think there is any kind of 'gap' between 'the mind' and 'the body.'

Notes

1. "Descartes had enough common sense to acknowledge the two-way interrelationships between mental and physiological functions, and enough intellectual honesty to recognize that he had a problem on his hands" (Wiredu 2004: 202).
2. The only *metaphysical* relation that can obtain between a thing and itself is identity: other relations, for example, 'linguistic' relations, can obtain between a thing and itself (for example, a thing can 'resemble' itself, be 'situated in the same office as' itself et cetera) only in virtue of *how those things are described*, by a linguistic agent.
3. Wiredu claims that dualism shares with materialism "the fundamental assumption … that the mind is a substance or [at least] *an entity* of some kind," which is "a conceptual inconsistency dear to much Western metaphysics" (2004: 204, my emphasis). This seems to be much the same as the claim that I am making here.
4. This is a fairly 'shallow' point: I do not intend to make any deep metaphysical claim about dualism/materialism, but rather to extend Papineau's point about 'closet' dualism. My claim is that it is *in virtue of the vicious reification of 'mind'* that one is able to diagnose 'closet dualism': if somebody commits vicious reification with regard to 'the mind' (which is to say, if she conceptualizes 'the mind' as if it were a 'thing,' comparable to material things), that is fairly conclusive evidence that the person must be a (closet) dualist.
5. Wiredu (1987: 168), like Ryle [Sect. 2.3.1], uses the term 'category mistake' to refer to this kind of usage.
6. Corcoran agrees: "Searle's biological naturalism is, despite Searle's protests to the contrary, a form of dualism" (2001: 307).

References

Boag, S. 2015. In defence [sic] of unconscious mentality. In *Psychoanalysis & philosophy of mind: Unconscious mentality in the 21st century*, ed. S. Boag, L.A.W. Brakel, and V. Talvitie, 239–265. ProQuest Ebook Central Database. https://0-ebookcentral-proquest-com.ujlink.uj.ac.za/lib/ujlink-ebooks/detail.action?docID=2082961&query=psychoanalysis+and+philosophy+of+mind. Accessed 10 September 2019.

Chalmers, D.J. 1995. *Facing up to the problem of consciousness*. http://consc.net/papers/facing.pdf. Accessed 23 March 2015. [Also published in *Journal of Consciousness Studies* 2 (3): 200–219].

Churchland, P.S. 1986. *Neurophilosophy: Toward a unified science of the mind-brain*. https://www.federaljack.com/ebooks/Consciousness%20Books%20Collection/Patricia%20Churchland%20-%20Neurophilosophy.pdf.

Churchland, P.M. 2013. *Matter and consciousness*. 3rd ed. MIT Press. https://epdf.tips/matter-and-consciousness-third-edition-2013.html.

Corcoran, K. 2001. The trouble with Searle's biological naturalism. *Erkenntnis* 55 (3): 307–324.

Cottingham, J. 1996. General introduction: The Meditations and Cartesian Philosophy. In *Meditations on first philosophy*, ed. J. Cottingham, Rev. ed. Cambridge University Press. https://doi.org/10.1017/CBO9780511805028.

Descartes, R. 1641/1996. *Meditations on first philosophy*. Trans. and Ed. J. Cottingham. Cambridge University Press. https://doi.org/10.1017/9781107416277.

Hennig, B. 2010. Consciousness as spontaneous knowledge. In *Essays on the concept of mind in early-modern philosophy*, ed. P. Glombiček and J. Hill, 7–27. ProQuest Ebook Central Database. https://0-ebookcentral-proquest-com.ujlink.uj.ac.za/lib/ujlink-ebooks/reader.action?docID=1114206. Accessed 19 June 2019.

Hill, J. 2010a. Introduction. In *Essays on the concept of mind in early-modern philosophy*, ed. P. Glombiček and J. Hill, 1–6. ProQuest Ebook Central Database. https://0-ebookcentral-proquest-com.ujlink.uj.ac.za/lib/ujlink-ebooks/reader.action?docID=1114206. Accessed 19 June 2019.

———. 2010b. Res cogitans as res dubitans. In *Essays on the concept of mind in early-modern philosophy*, ed. P. Glombiček and J. Hill, 29–43. ProQuest Ebook Central Database. https://0-ebookcentral-proquest-com.ujlink.uj.ac.za/lib/ujlink-ebooks/reader.action?docID=1114206. Accessed 19 June 2019.

Huxley, T.H. 1874. On the hypothesis that animals are automata, and its history. *Nature* 10 (253): 362–366.

Kwame, S. 2004. Quasi-materialism: A contemporary African philosophy of mind. In *A companion to African philosophy*, ed. K. Wiredu, 243–251. Wiley. https://studynotesunisa.co.za/wp-content/uploads/2019/01/blackwell-companion-to-african-philosophy.pdf. Accessed 20 January 2020.

Montemayor, C., and H.H. Haladjian. 2015. *Consciousness, attention, and conscious attention*. ProQuest Ebook Central Database. https://0-ebookcentral-proquest-com.ujlink.uj.ac.za.

Nagel, T. 1974. What is it like to be a bat? *The Philosophical Review* 83 (4): 435–450.

Panksepp, J. 2003. At the interface of the affective, behavioral, and cognitive neurosciences: Decoding the emotional feelings of the brain. *Brain and Cognition* 52: 4–14. https://doi.org/10.1016/S0278-2626(03)00003-4.

Papineau, D. 2002. *Thinking about consciousness*. https://www.google.co.za/url?sa=t&rct=j&q=&esrc=s&source=web&cd=4&cad=rja&uact=8&ved=0ahUKEwj5-9On9vHVAhXDJMAKHYs7DOQQFggzMAM&url=http%3A%2F%2Fwww.thebookishblog.com%2Fthinking-about-consciousness.pdf&usg=AFQjCNHhSO3QhdSfSt-w9gKUbMPggZ9VfA. Accessed 7 December 2016.

———. 2011. What exactly is the explanatory gap? *Philosophia* 39: 5–19.

———. 2020. Naturalism, in *The Stanford Encyclopedia of Philosophy*, edited by E.N. Zalta. https://plato.stanford.edu/archives/sum2020/entries/naturalism/

Pataki, T. 2000. Psychoanalysis, metaphor and the concept of mind. In *The analytic freud*, ed. Michael Levine, 47–136. London: Routledge.

Ryle, G. 1949. *The concept of mind*. London: Routledge. http://s-f-walker.org.uk/pubsebooks/pdfs/Gilbert_Ryle_The_Concept_of_Mind.pdf. Accessed 20 September 2015.

Schnaitter, R. 1985. The haunted clockwork: Reflections on Gilbert Ryle's The Concept of Mind. *Journal of the Experimental Analysis of Behavior* 43: 145–153. https://doi.org/10.1901/jeab.1985.43-145.

Searle, J.R. 1997. *The mystery of consciousness*. 7th ed. New York: NYREV.

Smithies, D. 2011. What is the role of consciousness in demonstrative thought? *Journal of Philosophy* 108 (1): 5–34. https://doi.org/10.5840/jphil201110811.

Stoljar, D. 2015. Physicalism. In *The Stanford encyclopedia of philosophy*, ed. E.N. Zalta. Stanford University Press. http://plato.stanford.edu/archives/spr2015/entries/physicalism/.

Tanney, J. 1949. Rethinking Ryle: A critical discussion of *The Concept of Mind*. In *The concept of mind*, ed. G. Ryle, ix–lvii. London: Routledge.
———. 2015. Gilbert Ryle, in T*he Stanford Encyclopedia of Philosophy* (Spring 2015 ed.), edited by E. N. Zalta. https://plato.stanford.edu/archives/spr2015/entries/ryle/.
Tye, M. 1999. Phenomenal consciousness: The explanatory gap as a cognitive illusion. *Mind* 108 (432): 705–725.
Van Riel, R., and R. Van Gulick. 2019. Scientific reduction. In *The Stanford encyclopedia of philosophy*, ed. E.N. Zalta. Stanford University Press. https://plato.stanford.edu/archives/spr2019/entries/scientific-reduction.
Von Neumann, J. 1958. *The computer and the brain*. New Haven: Yale University Press.
Wiredu, K. 1987. The concept of mind with particular reference to the language and thought of the Akans. In *Contemporary philosophy: A new survey*, ed. G. Fløistad, vol. 5, 153–179. Dordrecht: Martinus Nijhoff Publishers.
———. 1996. A philosophical perspective on the concept of human communication. In *Cultural universals and particulars: An African perspective*, ed. K. Wiredu, 199–206. Indiana University Press.
———. 2002. Conceptual decolonization as an imperative in contemporary African philosophy: Some personal reflections. *Rue Descartes* 2 (36): 53–64. https://doi.org/10.3917/rdes.036.0053.
———. 2004. Amo's critique of Descartes' philosophy of mind. In *A companion to African philosophy*, ed. K. Wiredu, 200–206. Wiley.

3

The Hard and the Easy Problems of Consciousness

I explained in the previous chapter why it was problematic to reify 'the mind': whenever a professed materialist resorts to talking about 'the mind' and 'the body,' the implication is that that 'materialist' is probably a closet dualist [Sect. 2.2.3]—and I stipulated that I will not consider dualism to be plausible. Similarly, it is problematic to reify 'consciousness': 'consciousness' is not an *entity* which exists in the (physical) world, but a *concept* which exists in language [compare: Wiredu 1987, Sect. 2.2.3]. In this case, vicious reification creates an opposition between 'consciousness' and 'the brain' which is very little better than the corresponding opposition between 'thinking things' (or 'minds') and 'physical bodies' [Sect. 2.2.1]. And just as in the case of 'the mind,' thinking of 'consciousness' in an overly concretized way makes it easy to forget that it is a kind of umbrella term, encompassing a specific group of experiences, and (mistakenly) to regard 'it' as having causal power [compare: Sect. 2.2.3.1]. (I wonder: might it even be the case that the use of 'consciousness' and 'the brain' is every bit as indicative of closet dualism as is the use of 'the mind' and 'the body'?)

© The Author(s), under exclusive license to Springer Nature Switzerland AG 2022

J. Anderson, *Biological Naturalism and the Mind-Body Problem*,
https://doi.org/10.1007/978-3-030-99684-0_3

In this chapter, I need to use the term 'consciousness' in order to discuss some of the so-called 'consciousness' research that has been done—but when I use this term, I will not be speaking *strictly*, hence the scare quotes.

3.1 Introduction: Investigating 'The Easy Problems'

As I explained [Sect. 2.3.3.1], Chalmers is well-known for his division of 'the problem of consciousness' into 'the hard problem' and 'the easy problems.' The easy problems were those that could be readily addressed using the methods of cognitive science, but the hard problem—namely, the problem of experience—resisted such methods. Chalmers said that, because the hard problem was so hard to investigate, a large part of 'consciousness' research turned out to be research into one of the easy problems, even when the target of the research was supposed to be phenomenal consciousness, in the 'hard-problem' sense—and my survey of the field in Sects. 3.2 and 3.3 certainly seems to bear out this suggestion.

For example: explaining the problem of the 'explanatory gap,' Ned Block takes 'the experience of green' as a paradigm instance of phenomenal (what-it-is-like) experience (2009: 1113). Admittedly, 'the experience of seeing green' comes a bit closer to the heart of Nagel's type of what-it-is-like experience than, for example, 'performing a mathematical calculation in one's head,' or 'being "conscious" of a certain word written on a page' et cetera; but there is no denying that out of all the possible subjective/first-personal/what-it-is-like *experiences* he could have used for his example, Block has not used an example like 'the experience of feeling happy,' or 'the experience of being very stressed before the exam.' He has unambiguously referred to one of 'the easy problems.'

In particular, it seems to me, 'consciousness' research often turns out to be research into the easy problems of un/conscious *sensory perception* (or cognition).[1] If there is a difference between 'consciously' perceiving something and 'unconsciously' perceiving something—or consciously or unconsciously performing some cognitive operation—it must, it is tempting to assume, be 'consciousness' itself which is present in one case, and absent in the other. Therefore, when I refer to 'problems to do with (conscious) sensory perception (and/or cognition),' I am largely referring to the same problems Chalmers refers to when he says 'the easy problems of consciousness'.[2]

But in spite of his recognition of the distinction between the easy problems of consciousness (in terms of sensory perception/of cognition), and the hard problem (of experience), I am concerned that even Chalmers himself is still thinking of 'the hard problem' in nearly the same way as he is thinking of the easy problems. He says that "A central fact about *experience* is that it has a complex structure. *The visual field*, has a complex geometry, for instance" (1995: 18, my emphasis). What I want to draw attention to here is that, for Chalmers, the complex geometry *of the visual field* is a good example of the complex structure *of experience*—which implies that he takes 'the visual field' and 'experience' to be very nearly interchangeable. But it was the *easy* problems of consciousness, and not the 'hard problem' of experience, which were supposedly defined as 'problems directly susceptible to the standard methods of cognitive science'; and in fact, Chalmers specifically mentioned that 'the ability to discriminate, categorize, and react to (visual) environmental stimuli' was an example of an easy problem [Sect. 2.3.3.1]. So now, when he says that the complex geometry of the visual field is a good example of the complex structure of (hard-problem) *experience*, he seems to be doing exactly the same thing he accused everyone else of doing: investigating the easy problems of consciousness (which I suggested could be understood as 'problems to do with sensory perception /with cognition'), while leaving the hard problem untouched.

Churchland's project in *Neurophilosophy* is also a good example of a philosopher's investigating the easy problems only, at the expense of the hard problem. She says that what we want to know is "how humans and other animals learn, what makes them smart, how they plan and problem solve, how they recollect and forget … what consciousness is." She admits that there is no particular reason for thinking that our word 'consciousness' picks out a single natural kind, or, on the contrary, whether it can be used to cover several different phenomena; but what we can be reasonably confident about is that when she says she wants to find out what 'consciousness' is, she is thinking primarily of a kind of 'consciousness' that is similar to planning, problem-solving, recollecting and forgetting. That is to say, she appears to be taking it for granted that (hard-problem) 'consciousness' is similar to planning, problem-solving, *and other such easy problems*.

As I mentioned [Sect. 2.3.4.2], Searle thought that, in order to get some insight into 'consciousness,' it would be sensible to investigate 'processes in the brain that are psychologically real but which have no conscious manifestation,' like *blindsight*, for example. He is evidently also taking it for granted that a good way of finding out about (hard-problem) 'consciousness' would be to examine cases of disrupted *visual perception*—in other words, to examine an example of 'consciousness' in the easy-problem sense.

It might be possible to acquit Ryle of focusing only on (the easy problems of) 'conscious' (sensory) perception and cognition: as I explained, he is not really interested in *anything to do with* 'mental processes' or the 'streams of consciousness' in which they may occur [Sect. 2.3.1.2]. He thinks this whole manner of speaking is both unnecessary and undesirable, and that we should understand statements about 'mental conduct' to be essentially nothing more than statements about observable behavior. Moreover, he ascribes to 'the official doctrine' the view that introspection is similar in kind to perception (it is like the perception of the 'inner' workings of the mind, rather than perception of the external environment)—and, as we know, Ryle has a low opinion of the official doctrine [Sect. 2.3.1.1]. However, as I am going to explain [Sect. 3.3.1.4] he still tends to think of the easy problems as more paradigmatic of 'consciousness' than anything else.

The first part of Chap. 3 [Sect. 3.2] concerns some typical 'consciousness' research: just the kind of research which focuses on what Chalmers [Sect. 2.3.3.1] described as 'the easy problems' of consciousness, and which I suggested could also be characterized as problems to do with sensory perception and cognition. But I explain that actually, the majority of perceptual and cognitive processes take place *without* conscious awareness. In other words: because sensory perception and cognition seem to be almost entirely independent of 'consciousness,' we should be careful not to conflate '*consciousness*' with '(conscious) sensory *perception* or *cognition*.'

In the second part of the chapter [Sect. 3.3], I will suggest that even when researchers explicitly direct their attention towards 'the hard problem' of what-it-is-like/experience, they still tend to focus on what-it-is-like *to perceive* stimuli, or *to perform cognitive operations*. But what I argued in Sect. 3.2 above also applies here: much perception and cognition takes place without being 'conscious.' The implication is that if somebody is not 'conscious' of perceiving stimuli, or performing cognitive operations, then there is also *nothing*-it-is-like to perceive those stimuli, or perform those operations. Therefore, we should be careful not to conflate (hard-problem) *what-it-is-like/experience* with (perceptual-cognitive) '*consciousness*'.

My consideration, in this section, of an alternative way of conceiving (sensory) perceptual experience—in terms of *attention*—also supports my suggestion in Sect. 3.2 that we should be careful not to conflate '*consciousness*' with 'conscious' *perception* or *cognition*.

3.2 The Easy Problems of Perception and Cognition

Although much 'consciousness' research actually turns out to concern the easy problems of ('conscious') sensory perception, actually, so far from taking perception as a paradigm example of 'consciousness,' we might do better to say that perception is, by definition, '*un*conscious.' This is because "perceptual activity is largely automatic and not under conscious or intentional control (the orange on the desk cannot be perceived as purple through an act of will)" (Bargh and Chartrand 1999: 465). In other words: sensory perception is involuntary; and since we are not introspectively aware—let alone in control—of the processes taking place when we perceive, there is a sense in which we could say that the actual act of perceiving is 'unconscious.'

Both specialists and laypersons tend to understand 'intellectual' (or cognitive) phenomena as being paradigmatic of 'mental' phenomena.

However, very much the same thing can be said about cognitive activity as about perceptual activity: in our day-to-day lives, we routinely carry out complex cognitive operations without having any conscious awareness that we are doing so. For example, every time somebody engages in a conversation with somebody else, "the listener is aware of the meanings of the words uttered by the speaker but not of the phonological and linguistic principles by which the meaning of the speaker's utterance is decoded" (Kihlstrom 1987: 1447). Thus, in a similar sense, the listener and speaker could be said to be using these phonological and linguistic principles 'unconsciously.'

In this section, I'm going to discuss two perceptual phenomena—blindsight [Sect. 3.2.1] and priming [Sect. 3.2.2]—which demonstrate the significant extent to which 'unconsciously perceived' stimuli can influence subsequent test performance, as well as other behavior. Priming and blindsight demonstrate that it is meaningful to say the stimuli are *perceived*, even though the research participants may not be *conscious* of perceiving these stimuli. The fact that (sensory) perception and consciousness can be dissociated supports my suggestion that we should be careful not to conflate (hard-problem) '*consciousness*' with (easy-problem) '(conscious) *perception*.'

Lastly, I will discuss the extent to which people can be 'unconscious' of the cognitive operations they perform when carrying out various tasks [Sect. 3.2.3]. The studies discussed demonstrate how we can perform quite complex cognitive operations without having any conscious awareness of what we are doing. The fact that cognitive processes are often not 'conscious' implies that we should also be careful not to conflate (hard-problem) 'consciousness' with (easy-problem) '(conscious) *cognition*.'

3.2.1 Blindsight

The term 'blindsight' was coined by Sanders et al. and Weiskrantz et al. (both in 1974), to refer to "the ability of patients with absolute, clinically established, visual field defects … to detect, localize, and discriminate visual stimuli despite being phenomenally visually unaware of them" (Cowey 2010: 3). In these cases, the outermost 'cortical' layer of tissue, in the 'occipital' region of the brain—which is responsible for almost all visual perception—can only perform part of its normal function; accordingly, blindsight patients only enjoy a limited part of normal visual experience. The necessary receptor cells in these patients' eyes continue to function normally, but their brains are not capable of processing the sensory input and using it to generate perceptual experience. The patients are not 'conscious' of being able to see (either 'at all,' or 'in a specific portion of their visual field'), and are unable to report details about objects presented to them. Not surprisingly, blindsight has attracted much attention not only from those concerned with the mechanisms of visual perception, but also investigators in the cognitive sciences who are interested in 'consciousness' (for example, Cowey and Stoerig 1995: 247).

3.2.1.1 Patient TN

Patient 'TN' remains "the only available case in the literature" (De Gelder et al. 2008: R1128) in which a patient suffered complete destruction of the visual area of the brain, but no damage to any other brain regions. This patient is, to all intents and purposes, completely blind, after two successive strokes completely destroyed the areas of his brain normally responsible for vision. However, he demonstrated spectacular navigational skills: the experimenters built an obstacle course in a long corridor (consisting of chairs/boxes et cetera scattered around), and TN managed to navigate it successfully without even once colliding with any of the obstacles.

Because such cases of blindsight, unaccompanied by other neurological deficits, are so rare in humans, much of the blindsight research has been conducted on primates whose brains have been deliberately lesioned in the relevant areas—which raises a concern about the extent to which the findings hold in human cases. However, Cowey and Stoerig specifically addressed this worry, and they confirmed that the findings of primate studies could safely be extrapolated to human cases. They taught their monkeys to press a certain light on the display when they detected a brief luminance target—and they found that monkeys with damage to part of their visual cortex were able to detect the stimulus reliably, even in their blind fields, just as blindsight patients are. But then the monkeys learned a second task. In this task, the monkeys were taught to press a certain permanent outline on the screen whenever the rest of the screen was empty of stimuli. After the monkeys had learned this second task, the researchers presented 'a brief luminance target' similar to the ones they had already used to demonstrate that the monkeys had some awareness of stimuli in the 'blind' part of their visual fields—*and the monkeys consistently pressed the outline to indicate that the screen was blank*. In other words, their behavior was sensitive to the presence or absence of a visual stimulus that they 'denied' being 'conscious' of seeing—which is exactly the same thing that happens in the case of blindsight patients. This indicates that primate research can legitimately be used to shed light on human blindsight.

3.2.1.2 HOT Theory

There are two main ways of accounting for blindsight: 'higher order' accounts [Sect. 3.2.1.1] and 'dual streams' accounts [Sect. 3.2.1.2]. According to Higher Order Thought theory, "what makes a mental state conscious is that it is the object of some kind of higher-order representation" (Gennaro 2004: 1). HOT theorists "often refer to the case of blindsight as evidence in support of their theory" (Gallagher and Zahavi 2008: 57), because their theoretical framework makes short work of the phenomenon: according to HOT theory, 'blindsight' is just like normal visual perception, except that the requisite 'higher order thought'—which is what normally allows one to be 'conscious' of these visual perceptions—is lacking. Blindsight is the basic kind of perception that everybody has, but in normal people, this 'blindsight' is brought into the sphere of 'consciousness' by a 'higher order thought.'

Block takes a higher-order-type view. He thinks that, although "[s]ome information about stimuli in the blind field is represented in the brains of blindsight patients," these patients "cannot harness ... information [about objects in their blind fields] in the service of action" (1995: 227)—which suggests that the function of 'consciousness'[3] is to enable us to somehow use information (represented in the brain) to guide action.

'Consciousness' embodies the 'higher order thought' which enables non-conscious visual perception (common to blindsighters and normals) to enter 'consciousness.' Similarly, Smithies (2011: 25) thinks that the only way to explain the difference between normal sight and blindsight must be to refer to the presence or absence of consciousness.

Searle evidently takes a 'higher order' view of blindsight because he said that 'a promising line of attack' for finding out 'how' the brain causes 'consciousness' was to focus on clinical cases 'where conscious experience is in some way disturbed or disrupted,' and that blindsight might be just such a case [Sect. 2.3.4.2]. He says "[i]f we knew how consciousness got into vision, we might be able to locate the specific mechanisms by which the brain causes consciousness, both for vision and for other forms of consciousness" (Searle 1997: 200)—but saying 'If we knew how consciousness got into vision, we might be able to locate the specific mechanisms by which the brain causes consciousness' sounds to me suspiciously like saying 'If we knew how pairhood got into left- and right-hand gloves, we might be able to locate the specific mechanisms by which the gloves cause pairhood.' The fact that the HOT theory of blindsight relies so heavily on vicious reification weighs strongly against this theory (even before anything has been said about whether this view is empirically supported).

On the higher order view,

> It is curious (or it must seem so to the HOT theorist) that the blindsight subject, who is certainly capable of higher-order thoughts, continues to complain that he cannot see the stimulus. Indeed, one might argue, this complaint is itself a HOT that is directed at the non-conscious sensory state, and yet, this does not make that sensory state conscious. … So the real question is: What is it about the non-conscious sensory state that prevents higher-order access to it? There must be something that differentiates it from a conscious sensory state, other than a HOT. (Gallagher and Zahavi 2008: 58)

But asking 'What is it about the non-conscious sensory state that prevents higher-order access to it?' sounds a lot like asking 'What is it about the non-pairhood state (of the individual left- and right-hand gloves) that prevents pairhood-order access to it?' This is an example of reification; and since this reification is involved in logical reasoning, it is vicious: it renders that reasoning invalid. Moreover, the reification of concepts like 'non-conscious sensory state' and 'higher order thought' implies the tacit endorsement of dualism [Sect. 2.2.3], which I have stipulated I will not consider as plausible. Accordingly, higher order accounts of blindsight are problematic.

3.2.1.3 Dual Streams Theory

Rather than saying that blindsight is just like normal vision, except that it lacks an additional higher order thought, a much more plausible way to account for the phenomenon is to say that 'object recognition' and 'spatial perception' are processed differently in the brain—which is what Dual Streams Theory proposes. Gerald Schneider's 1969 work with hamsters provided compelling evidence for the hypothesis that the superior colliculus is specialized for perceiving the spatial orientation of stimuli, while the primary (occipital) visual area is specialized for perceiving the visual details of stimuli.

First, the hamsters were trained to turn their head in the direction of a sunflower seed held in an experimenter's hand; subsequently, they had to run down one of two pathways, and go through a certain door to find a sunflower seed. In this second task, the hamsters were required to make simple visual discriminations such as distinguish between black and white doors or doors with vertical or horizontal stripes (Gazzaniga et al. 2002: 180), in order to identify which one concealed the seed.

Schneider found that the hamsters' performance on visual identification tasks (distinguishing the black/white doors with the horizontal/vertical stripes) was severely impaired when they had lesions in cortical regions of their brains, but not when they had collicular (that is, subcortical) lesions. On the other hand, their performance on the sunflower seed localization task was unimpaired in the presence of cortical lesions, but severely impaired in the collicular lesions case. The superior colliculus is specialized for spatial orientation, while the primary (occipital) visual area is specialized for object identification. According to this theory, blindsight is *not* a case of otherwise-normal vision which lacks 'the spotlight of consciousness' illuminating it: it is vision which is normal in regard to determining *the location* of the visual stimulus, but deficient in regard to determining *the visual details* of the visual stimulus.

In 1973, Walter Pohl confirmed the dual streams theory in his primate research. Pohl used two tasks: in the 'landmark discrimination' task, two 'wells' were recessed into the tabletop, and one of them would contain food. A small cylinder was placed next to the baited well, so the monkeys could learn, by trial and error, that they could always get food from the well near the cylinder. But just when they had learned this rule, the experimenters would switch the task, so that the food was reliably hidden in the well *furthest* from the cylinder (Gazzaniga et al. 2002: 196). In the 'object recognition' task, the procedure was exactly the same, except that in addition to the cylinder that was placed next to the baited food-well, a cube was also used, to mark the non-baited food-well.

Pohl found that if the parietal lobe was damaged, the monkeys were able to learn the first 'contingency' in the landmark discrimination task (that is, 'food is near to cylinder'), but failed to improve with each reversal of the rule (from 'food is near to cylinder' to 'food is far from cylinder'). Each time the rule was reversed, they were effectively learning the task again from scratch, rather than being able to apply more or less the same rule, and simply switching 'near to cylinder' to 'far from cylinder.' The likely explanation for this is that the monkeys had difficulty recognizing that the cylinder could be in different spatial relationships with the food. In other words, parietal lesions disrupted the 'where' pathway (ibid.).

If, on the other hand, the temporal lobe was damaged, the monkeys performed normally on the landmark discrimination task, but struggled with the object recognition task—they took longer both to learn the initial contingency, and to switch over when the contingency was reversed. Just the fact that a cube was present in the display, even though the position of the cube was irrelevant, made it much more difficult for them to learn the connection between the cylinder and the food. The likely explanation for this is that the monkeys had difficulty distinguishing the cube from the cylinder. In other words, temporal lesions "disrupted the 'what' pathway" (ibid.).

Whereas HOT theory takes blindsight as a case in which patients have 'vision,' but not '(visual) consciousness,' dual streams theory suggests that it might be a case in which patients are able to perceive either 'object details' or 'object location,' but not both. Accounting for blindsight in terms of the dual visual processing streams in the brain is more satisfactory than accounting for it in terms of Higher Order Thoughts (which requires dualism/vicious reification). Therefore, higher order accounts of blindsight are not only problematic in themselves: they also fail to demonstrate that blindsight has anything to do with 'consciousness.'

3.2.1.4 The Lesson (of 'Blindsight') for Philosophy

Obviously, there is at least a sense in which it is a contradiction in terms to say both that somebody is 'blind' and that s/he can respond (in any way) to visual stimuli. Cowey suggests that "[i]f blindsight had instead been called residual visual processing after destruction of primary visual cortex ... it is likely that much of its controversial nature would not have arisen" (2010: 3). Similarly, Merker says that 'blindsight' is "infelicitously named," seeing that "to a first approximation, patients do *not* see these stimuli" (2005: 100) to which they are nevertheless able to respond. According to these experts, blindsight should not be seen as a phenomenon involving (the hard problem of) 'consciousness,' but as a phenomenon involving only (the easy problem of) how different parts of the brain are involved in different aspects of visual perception.[4]

The case of blindsight shows that easy-problem '(conscious, visual) perception' is independent of—and therefore should not be conflated with—hard-problem 'consciousness.' Moreover, the fact that blindsight studies are so often taken to be relevant to 'consciousness' supports Chalmers' contention that even when people profess to be interested in what-it-is-like/experience, they often turn out to be investigating only the easy problems of ('conscious') sensory perception (and/or cognition) [Sect. 2.3.3.1].

3.2.2 Priming

Another case which shows that sensory perception can take place in the absence of 'consciousness' is *priming*—which refers to the subliminal activation of certain representations or associations immediately before a task is carried out. When somebody views a stimulus under conditions which prevent the subject from perceiving it consciously, that stimulus can still influence the subject's subsequent performance on various tasks, even when the subject cannot consciously identify it or even detect its presence. Although priming results were initially a bit controversial, the idea that non-reportable stimuli can significantly influence a broad range of processes is now well-established.

In 'direct' measures, it can be demonstrated that participants can explicitly detect, identify, and/or discriminate stimuli which they have perceived only subliminally. When an 'indirect' measure is used, the participant's subsequent test performance or behavior is shown to be influenced by the prime. An example of an indirect measure is the 'classic' experiment in which people were interviewed over the telephone, and asked to rate how good their life had been overall up until that point. Subjects were reliably shown to rate their lives as better overall when the weather was sunny during the interview (Bargh 2014: 34). What was even more remarkable was that if the telephone interviewers drew the subjects' attention to the weather before they did their interviews, this weather-related effect was no longer observed! In other words, the weather influenced the subjects' ratings of their lives *only* when they were *not (consciously) aware* of the weather.

In (indirect) tasks based on reading or word recognition, primed items are recognized faster, and more reliably, than unprimed ones. For example, if a certain word is presented, too briefly for the participant to be 'conscious' of having seen that word, the participant is still more likely to use that word, or related words, on subsequent tasks. In 'primes-to-behavior' studies, priming has been shown to influence the way the subject behaves after unconsciously perceiving the prime stimulus. For example, if a research participant is set some kind of word puzzle that involves words related to old age, that person subsequently tends to walk along the corridor at a slower pace than somebody who has solved puzzles involving other words (for example, Bargh et al. 1996, in Newell and Shanks 2014)—because he or she still has the idea of old age 'unconsciously' in mind.[5]

3.2.2.1 The Lexical-Decision Task

A large part of the research on priming has employed a procedure developed by Schvaneveldt and Meyer, which they termed the 'Lexical-Decision Task.' Using this procedure, these authors demonstrated that people were faster in deciding that a certain string of letters was a word (as opposed to a meaningless combination of letters) when the word was presented immediately after a related word. For example, when the word *nurse* is presented directly after the word *doctor*, it is recognized as being a word more quickly than when it is presented after the word *bread*.

1. In a variation on this format, Schvaneveldt and Meyer presented their subjects with three horizontal strings of letters simultaneously. The subjects had to say 'yes' if all three strings were actual words, and 'no' in all other cases. On the trials in which all three stimuli were actual words, the authors found that "[r]eaction time … averaged 90 ± 17 msec faster for stimuli that included two associated words and one unassociated word … than for stimuli with three unassociated words" (1973: 404).
2. In those trials where at least one of the letter strings was a nonword, "[w]hen two words were displayed above a nonword, RT was 71 ± 23 msec faster if the two words were associated" (1973: 405).
3. On the other hand, there was no significant association effect if the nonword was either in the middle or bottom position.
4. Finally, "[w]hen two nonwords were presented in the top and middle display positions, the mean RT was … faster than when a single nonword appeared in the top position" (1973: 407).

The authors then discuss what exactly these results can be taken to show—for example, whether their data can be taken as providing evidence for or against the 'location-shifting' and/or 'spreading-excitation' models of information retrieval—but I am not concerned with evaluating such models. As far as this section is concerned, the relevant point is purely that it is not necessary to be 'conscious' of a sensory perception before that perception can meaningfully influence subsequent task performance and/or behavior. In other words, sensory perception is independent of (and therefore should not be conflated with) 'consciousness.'

3.2.2.2 Semantic Resemblance

Another 'seminal' contribution to the early priming literature was made by Anthony J. Marcel. Marcel ran an experiment in which single words were briefly flashed on a screen, too briefly to be consciously perceived, and immediately followed by a neutral pattern. Participants then had to try and 'remember' which word they had just seen. Marcel found that

> A small but significant proportion of erroneous word responses, while showing little graphic or phonological relation to the stimulus, bore a striking semantic relationship to it. Thus *green* led to responses such as 'blue' and 'yellow', *queen* to 'king,' *apple* to 'orange,' *light* to 'dark,' *happy* to 'joy,' *clock* to 'time,' *chair* to 'table.' (Marcel 1983: 201)

That is to say, subjects appeared to be exhibiting knowledge of the stimulus words at a semantic level (since they could answer correctly more often than would be expected by chance) without being able to report any other characteristics of the stimulus words. They were able not only to *perceive* the stimuli non-consciously, but also to perform some level of cognitive analysis on the stimuli, similarly non-consciously. The semantic resemblance studies thus demonstrate that cognition is independent of (and therefore should not be conflated with) 'consciousness.'

During the early 1980s, several other authors also produced studies which seemed to confirm Marcel's results (although many, if not all, of these studies "suffered from serious methodological flaws," Kouider and Dehaene 2007: 860). During the late 1980s, research on subliminal ('unconscious') perception entered a phase which Kouider and Dehaene characterize as 'Methodological improvements and recovery from scepticism' (2007: 861), and there was a renewed interest in priming phenomena in the field of psycholinguistics, as well as in scientific circles. By the end of the 1990s, not only had earlier results been replicated, but the improved methodologies employed in priming studies made these results more credible than ever before. Although there is still some disagreement about exactly how subliminal stimuli manage to affect subsequent perception or task performance, "the existence of subliminal perception is largely acknowledged" (Kouider and Dehaene 2007: 863).

3.2.2.3 The Lesson (of 'Priming') for Philosophy

Newell and Shanks describe subliminal perception as "the controversial phenomenon whereby invisible stimuli may influence some aspect of behavior" (2014: 13; see also Kouider and Dehaene 2007: 362)—but this seems to me unnecessarily counterintuitive. Obviously, if something is literally 'invisible' to someone, then it cannot influence anything else that person does. I would put it like this: 'under certain conditions, a stimulus that is not perceived *consciously* is still *perceived*, and therefore may influence a subject's subsequent sensory perceptions, cognitions, or behavior.' The case of subliminal priming therefore indicates that perception is independent of 'consciousness.' Moreover, a stimulus of which somebody is not 'conscious' can still influence that person's subsequent cognitive processes quite strongly—without that person's being aware that such an influence is at work. Therefore, priming shows that sensory perception and cognition are both independent of (and therefore should not be conflated with) 'consciousness.'

Moreover, we should bear in mind that investigating priming is a case of investigating un-/conscious perception and cognition—which supports Chalmers' claim that much of the so-called 'consciousness' research actually turns out to be research into 'the easy problems'—which I claimed were largely problems to do with sensory perception and with cognition, Sect. 3.1.

3.2.3 Introspective Access to Cognitive Processes

As we saw [Sect. 2.2.1.1], Descartes had a strong conviction that living bodies—including human bodies—operated essentially automatically (without 'volition' or 'conscious will'). Most typical twentieth-century research suggests that, similarly, much of our cognitive processing is 'unconscious' or automatic (for example, Dresp-Langley 2012: 2). *Telling more than we can know: verbal reports on mental processes*, by Nisbett and Wilson, was very influential among psychologists in the 1970s, and even now, "[t]o a considerable extent, the willingness of contemporary experimental psychologists to embrace the possibility of unconscious influences on behavior can be traced to the highly influential work of Nisbett and Wilson (1977)" (Newell and Shanks 2014: 4). Therefore, I am going to examine this study in some detail. In the following section [Sect. 3.2.3.2], I will examine Newell and Shanks' response to Nisbett and Wilson's article, in similar detail.

3.2.3.1 Telling More than We Can Know: Verbal Reports on Mental Processes (Nisbett and Wilson 1977)

The first part of the paper comprises a review of studies on subjective reports on mental processes. The authors claim that the literature they reviewed shows that people not only

1. may have very limited ability to report accurately on the cognitive processes that they are using during task performance, but also
2. may be completely unaware of perceiving stimuli that are influencing their performance (as in the case of subliminal perception)—they even
3. may be unaware that any kind of perceptual and/or cognitive process is taking place, during task performance, at all.

The authors are not going quite so far as to say that people *do not ever* have introspective access to their own cognitive processes, but they certainly seem to be tending towards this view. In the same way Bargh and Chartrand said that perception was unconscious, almost by definition (it not being possible, for example, to perceive the orange on the desk as purple through an act of will, Sect. 2.2), so we might be justified in saying that basic cognitive processes were necessarily unconscious. “It would be absurd, for example, to ask a subject … whether he stored the meanings of animal names in a hierarchical tree fashion or in some other manner” (Nisbett and Wilson 1977: 232).

3.2.3.1.1 Cognitive Dissonance

A large part of the evidence which purportedly shows that people are ‘unconscious’ of the cognitive operations they are performing comes from the literature on ‘cognitive dissonance’ studies—the central idea of which is that

> behavior that is intrinsically undesirable will, when performed for inadequate extrinsic reasons, be seen as more attractive than when performed for adequate extrinsic reasons … [A]ttitude change occurs because the cognition ‘I have done something unpleasant without adequate justification’ is dissonant and therefore painful; and the person revises his opinion about the behavior in order to avoid the psychic discomfort. (Nisbett and Wilson 1977: 233)

In a 1969 study, researchers asked their participants if they would be willing—as part of a certain research project—to undergo painful electric shocks whilst performing a certain learning task. Then they asked the subjects to go through the same shocks a second time. The participants were given either a good or a bad reason to justify undergoing the second round of shocks (“the research was very important, nothing could be learned unless the shocks were given again,” or “the experimenter wanted to satisfy his more or less idle curiosity about what would happen,” Nisbett and Wilson 1977: 234)—and the result was that those who had been given sufficient justification for going through the unpleasant process actually did worse on the learning task than those who had been given insufficient justification. This was explained by proposing that the group with insufficient justification, having decided to undergo the shocks, needed to find a way of justifying to themselves the fact that they were going to do something so clearly *not* in their own best interests—and so they 'determined' that the shocks were actually "not all that painful" (ibid.). The physiological (heart rate) and behavioral (test performance) indicators of pain displayed by these participants indicated that they really did experience lower stress levels than the other group, *purely because they had this (unconscious) reason for wanting to think that the shocks were not very painful.*

A similar study showed slides of snakes to snake-phobic subjects while administering occasional electric shocks to them as well. The subjects were fitted with wiring which they were told would record their heart rates while they viewed the slides, and they heard "a rhythmic pattern of sounds which, they were told, was the amplified sound of their own heart beats" (Nisbett and Wilson 1977: 234)—but in fact they were just being played a random pattern of sounds which had nothing to do with their hearts. Among the slides of snakes there were some slides printed with the word SHOCK, and each time this slide was shown, the subjects would be given an electric shock. After a few trials, the investigators started to play the rhythmic sounds at an increased rate of speed whenever the SHOCK slide appeared—but never when the snake slides appeared. Thus the subjects were led to believe that they were a little nervous of electric shocks (as measured by their anticipatory increased heart rate), but hardly nervous of snakes at all. Following this procedure, these subjects were subsequently brave enough to go much closer to a laboratory boa constrictor than the controls, who had not been led to believe that the rhythmic sounds were their own heartbeats.

Even though the subjects were not 'conscious' of the inferences they were drawing, they were still using 'cognition' (or logic) correctly to predict how their bodies would respond if they were or were not feeling anxious. The study presumably would not have worked on infants, for example, because they would have lacked the understanding necessary to make the logical connection between getting the shocks, or seeing the snakes, feeling anxious, and therefore experiencing an increased heart rate. This indicates not only that logical inferences can be drawn without 'consciousness' being 'present,' but also that the subjects' (unconscious) cognitive beliefs could significantly influence their (conscious) what-it-is-like/experience. That supports my claim that cognition (in the form of drawing correct logical inferences) is independent of (and therefore should not be conflated with) 'consciousness.'

3.2.3.2 Unconscious Influences on Decision-Making: A Critical Review (Newell and Shanks 2014)

Newell and Shanks first discuss Nisbett and Wilson's paper in some depth, because Nisbett and Wilson's work was highly influential in persuading people that much cognition was not conscious [Sect. 2.2.3.1]; then they also did some studies of their own.

3.2.3.2.1 Deliberation Without Attention

"[T]he idea that conscious deliberation is the ideal (if not always attainable) way to approach a decision forms the background of classic ... as well as contemporary decision making" (Dijksterhuis et al. 2006: 1005). On the other hand, "the notion that unconscious thought is fruitful hardly developed beyond the status of 'folk wisdom'" (that is, we have some idea that 'sleeping on it' is a useful approach to deciding what to do in a difficult situation, but this idea is not the result of any *theoretical* considerations: it comes from the *practical observation* that this course of action can be beneficial). But Dijksterhuis and his group have produced several studies which suggest that when we are faced with 'complex decisions,' the best thing to do (as measured by 'post-choice satisfaction' with our decisions) is "to stop thinking and let our unconscious decide" (Newell and Shanks 2014: 8). Although there is some concern about the 'robustness' of this finding, it is generally accepted that the deliberation-without-attention effect is to at least some extent supported by empirical evidence.

A standard experiment for comparing 'conscious' and 'unconscious' forms of deliberation is something like the following:

> [P]articipants are presented with information about three or four objects (e.g., cars) described by 10 or more attributes (e.g., mileage) and are asked

> to choose the best object. In most experiments *best* is determined normatively by the experimenter assigning different numbers of positive and negative attributes to each option. Attribute information is presented sequentially and typically in random order about the four options. Following presentation of the attributes, participants are assigned to one of three (or sometimes only two) conditions. In the unconscious thought condition, participants are prevented from making decisions for a few minutes by engaging in some distracting activity (e.g., solving anagrams). This distraction period is claimed to facilitate unconscious thought—'cognitive and/or affective[6] task-relevant processes [which] take place outside of conscious awareness' (Dijksterhuis 2004, p. 586). In the conscious thought condition participants are asked to think carefully about their choice for a few minutes, while in the immediate condition participants are simply asked to make their decision as soon as the presentation phase has finished. (Newell and Shanks 2014: 8)

For example, in one study, participants were told to choose their favourite poster from five posters, and afterwards were questioned about their satisfaction with their choice. "[P]articipants who chose their favorite poster … after thorough contemplation showed less postchoice satisfaction than participants who only looked at them briefly" (Dijksterhuis et al. 2006: 1005). The authors give two reasons for these results: first, the low capacity of 'consciousness' means that conscious choice-makers are unable to consider all the relevant factors in their choice; secondly, conscious thought about choices allows our biases to come into play, because individuals care more about some elements of the situation/context than others, and this prevents them from being able to weigh all the options fairly against one another. Unconscious thought processes do not face these obstacles, and can lead to better choices in certain situations.

While Newell and Shanks admit that the Dijksterhuis-type experiments seem, "to some extent, to provide empirical evidence for the soundness of the deliberation-without-attention recommendation" (2014: 10), they have concerns about the validity and reliability of the studies, and suggest that we should therefore continue to be wary of being too credulous. "[E]ven if the [deliberation-without-attention] effect can be reliably obtained, its existence falls well short of providing unequivocal evidence for the involvement of active unconscious processes" (ibid.). But I am not concerned with the design of the studies, or with whether they are sufficiently rigorous to provide *unequivocal* evidence for their claims. All I am interested in is the idea that a period of post-choice distraction *quite reliably* seems to lead to better choices than either conscious thought or an immediate decision—and this seems to be more or less uncontroversial. If this is indeed the case, this also supports my claim that cognition (in the form of making informed choices) is independent of (and therefore should not be conflated with) 'consciousness.'

3.2.3.3 The Lesson (of 'Introspective Access to Cognitive Processes') for Philosophy

Newell and Shanks expressed concerns about the way in which the deliberation-without-attention studies were designed and implemented (and therefore the validity/reliability of these studies); however, they did not deny that the deliberation-without-attention effect probably does exist. Therefore, Newell and Shanks's criticisms notwithstanding, this research paradigm seems to provide quite substantial support for the idea that cognition may be independent of 'consciousness.' Moreover, in the Cognitive Dissonance studies, Nisbett and Wilson found that when participants were *forced or bribed* into writing an essay advocating views which conflicted with their own views, their attitudes did not change; however, when they were given *a poor* reason for writing the essay, their views shifted toward the position advocated in the essay. The claim was that, when they only had a poor reason, these participants *unconsciously* changed their beliefs, in order that their beliefs would no longer conflict with the views they were expressing in the essay. To me, that seems like a good point; if that had been the only point that Nisbett and Wilson had made in their paper, that would have been enough for me to consider the existence and/or prevalence of unconscious cognitive processes to have been quite convincingly demonstrated. And this is a point about which Newell and Shanks appear unable to make any complaints.

According to Drew Westen, "the proposition that many cognitive processes are carried out unconsciously is as widely accepted by experimental psychologists as the opposite proposition was by their predecessors less than 2 decades ago" (1998: 336). In other words: cognition is independent of—and therefore should not be conflated with—'consciousness.'

3.2.4 The Easy Problems of Perception and Cognition: Conclusion

In this section [Sect. 3.2], I have shown that, quite apart from the metaphysical problem inherent in trying to account for thinking/consciousness within either dualist or 'materialist' frameworks [Sect. 2.2.3], there are some serious problems with attempting to understand 'consciousness' by means of studying ('conscious') sensory perception or cognition.

The only time the phenomenon of blindsight can be taken to have anything at all to do with 'consciousness' is when one takes a 'higher order' view of vision, according to which 'conscious vision' comprises 'unconscious-vision-plus-consciousness.' But higher order accounts are not plausible. On alternative accounts, such as the 'dual streams' account, the phenomenon of blindsight fails to demonstrate that (sensory) perception is dependent on 'consciousness.'

Subliminally-perceived stimuli influence subsequent perceptual and cognitive processes, in spite of the fact that participants are not 'conscious' of these stimuli. Priming therefore shows that both perception and cognition are independent of (and therefore should not be conflated with) 'consciousness.'

Regardless of what we think about whether specific studies have been well designed or poorly designed, we can be reasonably confident that there are several types of cognitive processes which are completely independent of (and therefore should not be conflated with) 'consciousness.' The cognitive dissonance studies seemed to provide particularly compelling evidence in this regard. Therefore, because perception and cognition seem to be almost entirely independent of 'consciousness,' we should be careful not to conflate *consciousness* with ('conscious') perception or cognition.

In addition, "[s]ince there can be sensation without the capacity for thought, as in the more lowly animals, it seems philosophically judicious to keep the distinction between the two types of phenomena always clearly in mind" (Wiredu 1987: 172), especially when one is specifically interested in *human* 'consciousness.' There is also a "categorical disparity" between sensation and thought, seeing that "sensations can be located in space (portions of skin, for instance) in a way which is inapplicable to thoughts" (ibid.). Finally: "sensations are ... incapable of referring to anything beyond themselves, whereas this referential capacity is the very defining characteristic of thought." Wiredu thinks that "[t]he categorical separation ... between mind and sensation seems ... apt," which is a view I fully endorse (except we must remember that Wiredu can be using the word 'mind' only in its 'scare-quotes' sense; compare: Sect. 2.1).

3.3 The Hard Problem of What-it-is-like/experience

Since perceptual and cognitive processes seem to be almost entirely independent of 'consciousness' (as I explained in Sect. 3.2), we should be careful not to conflate *consciousness* with (sensory) perception or cognition. But this also entails that perception and cognition should not be conflated with *what-it-is-like/experience*. If perceptual or cognitive operations take place 'unconsciously,' then there is nothing-it-is-like to perform those perceptual or cognitive operations (Chalmers, in Sect. 2.3.3.1).

We have already seen above [Sect. 3.1] how several investigators take various types of (conscious) perception (especially *visual* perception), and/or of (conscious) cognition, to be paradigmatic of (phenomenal) 'consciousness' in general. In Sect. 3.3 I will first give some examples of my four philosophers (from Sect. 2.3) discussing examples of 'consciousness' that have less to do with what-it-is-like *to perceive* a certain thing, or perform a certain *cognitive* operation (that is, the easy-problem sense of consciousness), and more to do with what-it-is-like *to be* a certain way (in the hard-problem sense) [Sect. 3.3.1]. However, I am claiming that even when these authors manage to come up with non-perceptual-cognitive (that is, non-easy-problem) examples of 'consciousness,' they still take the perceptual-cognitive examples to be the most paradigmatic ones. Secondly [Sect. 3.3.2], I am going to suggest that it is quite satisfactory to think of these cases in terms of whether or not we are *paying attention* to our perceptions/cognitive operations, rather than whether or not we are 'conscious' of them. This is another reason for us to be cautious of conflating 'consciousness' with ('conscious') perception or cognition.

3.3.1 What-It-Is-Like in Philosophy of Mind

3.3.1.1 Chalmers

I have already explained [Sect. 3.1] that Chalmers appeared to take 'the structure of the visual field' to be virtually identical to 'the structure of experience' (which entails that he also takes 'visual perception' to be virtually identical to 'experience'); I claimed that, because he focuses on visual consciousness, he can be thought of as taking an overly perceptual view. Chalmers even thinks that a good example of a hard-problem question is "Why is it that when our cognitive systems engage in visual and auditory information-processing, we have visual or auditory experience?" (1995: 3): on his view, clearly, somebody's being 'conscious' of something is best described as the case in which a *cognitive system* 'processes' information. Therefore, it seems fair to say that he also takes an overly *cognitive* view. In spite of his complaint that most consciousness research left the hard problem untouched [Sect. 3.3.1.1], he himself seems to fall into the trap of conflating (hard-problem) consciousness with (easy-problem) 'conscious *perception* (and/or cognition).'

Even when he claims that he is talking "only about the subjective quality of experience: what it is like to be a cognitive agent" (1996: 5), we can see that, in one and the same sentence, he has both claimed that his concern is with what-it-is-like/experience, and revealed that he thinks that what-it-is-like *to be a 'cognitive agent'* is the relevant what-it-is-like/experience in the case of human persons. However, this is problematic: the implication is that Chalmers is only talking about 'the hard problem of consciousness' insofar as it applies to intellectually and/or cognitively mature adult humans. For him, it sounds as if any individual who lacked the requisite *cognitive capacity* might not enjoy any hard-problem consciousness (/what-it-is-like/experience) at all. This belies his claim that the kind of phenomenal what-it-is-like/experience he is interested in (that is, the hard-problem sense of consciousness) is different from the 'easy problem' sense of consciousness, which includes various cognitive abilities. Therefore, Chalmers' view of what-it-is-like/experience is problematic.

3.3.1.2 Churchland

Churchland also seems to be predominantly interested in various perceptual/cognitive (and especially, computational) capacities—all of which, I suggested, should be taken as 'easy-problem' consciousness phenomena [Sect. 3.1]. She does also admit that part of 'what we want to know' is "how [people's] brains differ when their personalities differ, how they can be self-aware, what consciousness is" (1986: 152), but the fact that she wants to situate these kinds of 'problems' within any kind of computational or information-processing framework [Sect. 2.3.2.2] suggests that she is still thinking more of (easy) problems to do with ('conscious') perception and/or cognition (or computation), rather than the (hard) problem of what-it-is-like/experience. That is to say, Churchland's view of what-it-is-like/experience is problematic.[7]

3.3.1.3 Searle

As we saw, Searle was interested in the problem of discovering a causal relationship between 'consciousness' and 'brain processes' [Sect. 2.3.4.1]. 'Conscious states' he took to be 'qualitative' "in the sense that for any conscious state … there is something that it qualitatively feels like to be in that state" (1997: xiii)—which looks like a good step in the direction of investigating (the hard problem of) what-it-is-like/experience rather than (the easy problems of) perception and cognition. But then Searle continues: "[t]he enormous variety of stimuli that affect us—for example, when we taste wine, look at the sky, smell a rose, listen to a concert ... cause ... awareness or sentience" (1997: 3). The point to notice here is that even when he says he is thinking about 'awareness or sentience' (in the hard-problem sense), he is clearly thinking mainly about awareness *of tasting*, *looking*, *smelling* and *listening to* things—where tasting, looking, smelling and listening to are all very clear cases of *sensory* (/perceptual/easy-problem) experiences.

Searle does go on to say that 'the problem' of how the brain causes consciousness "is *not just* about the perceptual cases I have mentioned, *but includes* the experiences of voluntary actions, as well as such inner processes as worrying about income taxes or trying to remember your mother-in-law's phone number" (1997: 3, my emphasis). We experience "bodily sensations such as pains in the knee or the taste of honey" (and "visual perceptions such as seeing a rose"); experiencing 'consciousness' "*also includes* thinking about mathematical problems or the next election" (1997: 200–201, my emphasis)—but I think it is reasonably clear that for Searle, the first, most paradigmatic and most uncontroversial examples are the perceptual-cognitive ones. The fact that Searle says the other examples are *also* examples of experiencing consciousness shows that he takes them to be different from—and less central than—the first, perceptual-cognitive examples. Therefore, Searle's view of what-it-is-like/experience is problematic.

3.3.1.4 Ryle/Tanney

Ryle suggested that the best way to avoid making a category error would be to give up the use of mental conduct verbs altogether, and acknowledge that the only thing we are qualified to talk about is overt behavior. However, there are certain kinds of 'mental conduct' which are very difficult to talk about in this way. Ryle uses "silently humming a tune to oneself" and "one's description of last night's dreams" (Tanney 1949: xv) as examples of 'consciousness' in its hard-problem sense, which means that we can probably acquit Ryle of thinking of what-it-is-like *only* in perceptual/cognitive terms. But what Tanney says about Ryle is that "[a]lthough [he] does not deny—indeed he frequently peppers his discussions with—such episodes as hummings or dreaming, he seems to many … to go too far in *The Concept of Mind* to *minimize or downplay* their existence" (Ryle 1949: xv, my emphasis). And since using perceptual-cognitive phenomena as examples of 'hard-problem' consciousness is problematic, so minimizing or downplaying *non*-perceptual-cognitive examples of (hard-problem) 'consciousness' must be, to the same degree, also problematic.

3.3.1.5 The Lesson (of 'What-It-Is-Like in Philosophy of Mind') for Philosophy

I agreed with Chalmers when he said that most work on 'consciousness' leaves the hardest problems about consciousness untouched [Sect. 2.3.3.1]; in this section, I have demonstrated that even when they profess to be investigating 'the hard problem' of consciousness, investigators still commonly focus on what it is like *to experience various perceptual stimuli*, and *to perform various cognitive operations*, which is hardly different from investigating 'the easy problems' (which I equated with 'problems to do with perception or cognition,' Sect. 3.1). In contrast, Nagel's challenge for us was to imagine what it would be like *to be* a bat. This important kind of 'what-it-is-like' phenomenal consciousness (which is independent of conscious *perceptual* experience) seems to be widely 'downplayed' not only by Ryle, but by many other researchers in philosophy, as well as other mind sciences.

When some perceptual or cognitive process is not 'conscious,' that process cannot produce any what-it-is-like/experience. Therefore, supposed accounts of (hard-problem) 'consciousness' which are couched in terms of (conscious) sensory perception, or of cognition, are problematic, because, as I explained at some length [Sect. 3.2], both sensory perception and cognition seem to be almost completely independent of 'consciousness.' I am claiming that investigating perceptual and cognitive processes is just as unsatisfactory as a means of finding out about (hard-problem) *what-it-is-like/experience* [Sect. 3.3] as it is unsatisfactory as a means of finding out about (easy-problem) '*consciousness*.'

3.3.2 'Consciousness' vs Attention

'Attention' can be understood as "a selective processing mechanism ... that enhances and selects perceptual information for executing actions and higher-level cognition" (Montemayor and Haladjian 2015: 25). Therefore, attention investigators can account for whether or not one has some kind of what-it-is-like/experience not in terms of whether or not the experience is '*conscious*,' but rather, whether or not one is paying *attention* to what one is perceiving. The immediate advantage of this way of speaking is that it does not commit one to the (closet) dualism I said was implied by talking about 'the mind,' or 'consciousness' [Sect. 2.2.3]. Moreover, there seem to be some quite strong empirical considerations in favour of dissociating 'consciousness' from 'attention,' as I will now explain.

In the 1950s, it was discovered that air-traffic controllers, for example, were surprisingly bad at dealing with simultaneous incoming auditory messages; and even when subjects were evaluated on dichotic listening tasks (where they had to listen simultaneously to different messages coming in to their right and left ears), in the absence of all the stress of being an air traffic controller, they performed similarly badly (Kahneman and Treisman 1984: 29). This was explained by saying that people only had limited ('conscious') attentional resources, which were unable to cope with simultaneous competing demands.

In contrast, subsequent studies on attention suggested that (at least in the simpler experimental situations) whether or not someone was 'conscious' of a sensory perception could be completely independent of whether *attention* was being paid to the perception. According to Montemayor and Haladjian (2015: 3–4), views on the relation between 'consciousness' and 'attention' now fall along a spectrum, ranging from views that actually conflate consciousness and attention (for example, Prinz 2012), via views that suggest consciousness and attention are often dissociated, (for example, Koch and Tsuchiya 2007), to the view that consciousness and attention are completely separate, and do not overlap at all (for example, Tallon-Baudry 2012). Montemayor and Haladjian's own view (2015: ix) is that, although it seems intuitive to identify 'attention' with 'consciousness,' the concepts are actually independent of each other, although there is a certain amount of overlap between them.

3.3.2.1 The Filtering Paradigm and the Selective Set Paradigm

Just like any other capacity of any organism, the capacity for selectively attending to stimuli must in some respect have been beneficial to those organisms in their struggle for survival, otherwise that capacity would not have been maintained through successive generations. There are two main schools of thought on the benefits of selective attention (Kahneman and Treisman 1984: 29). According to the first, attention is mainly beneficial insofar as it facilitates the processing of whichever sensory information is currently the most important, whilst successfully resisting distraction by less relevant stimuli. This is 'the filtering paradigm,' developed by Donald Broadbent, in an attempt to "model the ability to attend to a specific sequence of sounds amid a variety of auditory inputs" (Montemayor and Haladjian 2015: 44).

In the 'filtering' paradigm,

> (1) the subject is exposed simultaneously to relevant and irrelevant stimuli, (2) the relevant stimuli control a relatively complex process of response selection and execution, and (3) the property that distinguishes the relevant from the irrelevant stimuli is usually a simple physical feature and is different from the property that determines the appropriate response. (Kahneman and Treisman 1984: 31)

Colin Cherry (1953) invented the 'selective shadowing' task, in which participants had to 'shadow' (that is, repeat aloud) auditory input from one ear only, while ignoring simultaneous auditory input from the other ear. The results of this study indicate that very little information seemed to be obtained from the non-attended channel. This finding is often informally referred to as 'the cocktail party effect'—because somebody in a social environment (like a cocktail party) with lots of different conversations taking place simultaneously is usually quite successful in 'tuning out' all except the most relevant inputs. For example, if somebody says your name, that usually 'jumps out' from all the other conversation taking place in the background. This means that, even when people are not 'conscious' of doing so, they are still able to scan the auditory environment for any particularly relevant sounds.

Sperling's (1960) research on short-term visual storage, using the 'partial-report technique,' is also an example of the filtering paradigm. In Sperling's experiments, cards, with groups of letters and/or numbers on them, were briefly displayed in front of the participants, who subsequently had to try and write down the letters/numbers on an answer grid. In the 'partial report' condition, the subjects were only required to focus on a single row of characters—which row it was indicated by a tone, sounded immediately after the visual presentation of that relevant row. The idea behind this is that, in just the same way that students' knowledge of a subject can be approximately determined by testing them on a selection of questions, so subjects' knowledge (or recall) of a range of stimuli could be approximately determined by testing them on only a selection of the stimuli they have seen.

Sperling found that "the accuracy of partial reports was consistently greater than the accuracy of whole reports" (1960: 19). This can be explained by appealing to the fact that very short-term visual memory is capable of storing the information for only a very short time before it begins to fade. Therefore, when attention is focused only on a very limited aspect of this sensory information, there is a much better chance that the subject will be able to access the relevant information before it deteriorates. Focused attention is what makes the difference between being able to recall the relevant row rather well and rather poorly.

In both these types of study (that is, 'selective shadowing' and 'partial-report') in the 'filtering' paradigm, subjects do not appear to experience the irrelevant stimuli as highly distracting, and are still able to focus attention efficiently on the relevant stimuli. In other words, they can distinguish between relevant and irrelevant stimuli even if they are not 'conscious' of the stimuli. These findings thus confirm that attention is independent of 'consciousness.'

According to the second school of thought, the most important function of attention is to facilitate the adoption of the appropriate set of behavioral response tendencies. In the 'selective set' paradigm,

> the subject is prepared for particular stimuli and is instructed to indicate by a speeded response the detection or recognition of those stimuli. Thus, the subject chooses which of several *possible* stimuli to expect or search for rather than which of several *actual* stimuli to analyze. (Kahneman and Treisman 1984: 31)

A good example of selective set paradigm tests is studies of 'search' (that is, looking for a particular visual stimulus within an array of irrelevant visual stimuli). Before the subject performs the task, her attention is directed towards one or more specific targets. In these kinds of tests, the target stimuli appear to 'pop out' from among a few or even many distracting stimuli. This suggests that subjects fully process only those stimuli which they have determined in advance are relevant, and their processing of these stimuli is not much affected by the irrelevant stimuli. Priming studies [Sect. 3.2.2] are also an example of the selective-set paradigm, insofar as the primes function as cues which focus participants' attention towards those specific stimuli—and the effectiveness of this focused attention can be seen in the fact that primed stimuli can be identified more quickly than unprimed stimuli.

Just as I said I was not concerned with whether those priming studies I discussed supported the 'location-shifting' and/or 'spreading-excitation' models of information retrieval [Sect. 3.2.2.1], here, I am not concerned with exactly what the filtering paradigm and the selective set paradigm show regarding the correct way to conceptualize the mechanism of visual attention. All I am concerned with is the fact that all the authors I have mentioned are able to account for their findings satisfactorily within the framework of 'attention,' without referring to whether or not 'consciousness' 'accompanies' certain sensory perceptions. In other words, 'attention' is independent of (and therefore should not be conflated with) 'consciousness.'

3.3.2.2 Attention and Priming

As I mentioned [Sect. 3.2.2], priming studies are usually produced as evidence that stimuli which are processed 'unconsciously' still have the power to affect performance; but Kahneman and Treisman point out that "demonstrated subliminal effects indicate a dissociation between perception and consciousness that is not necessarily equivalent to a dissociation between perception and attention" (Kahneman and Treisman 1984: 31). In other words, even if it can be demonstrated that stimuli can be perceived without the subject's being *conscious* of perceiving those stimuli, that is not the same thing as demonstrating that stimuli can be perceived without the subject's *paying attention* to (the location of) those stimuli. During priming studies, participants are instructed to focus their attention on some neutral stimulus, referred to as the 'mask,' at the start of the procedure. The experimenter then tests whether the participants respond in the same way to subsequent stimuli when they are shown the mask only, and when they are shown the mask, followed by the subliminally-presented target stimulus. The mask has nothing to do with anything, but because they are told to focus on the mask before the primed stimulus is presented, participants are at the least likely to be *paying attention* to the location where the subliminal stimulus is subsequently shown.

A number of studies (reviewed by Kouider and Dehaene) all suggest that attention aids in the perception of all kinds of subliminal stimuli. If this is indeed the case, it might be possible to account even for priming effects quite satisfactorily in terms of perceptual attention, without having recourse to the concept of 'consciousness' at all. This remains "an open question" which has not been thoroughly investigated (2007: 867).

3.3.2.3 The Lesson (of 'Consciousness vs Attention') for Philosophy

As Kihlstrom said (already in 1987), "consciousness is not to be identified with any particular perceptual-cognitive functions [since] these functions can take place outside of phenomenal awareness. Rather, consciousness is an experiential quality that may accompany any of these functions"

(1987: 1450). Seeing that I have argued [Sect. 2.2.2.4] that the use of terms such as 'the mind' (or 'consciousness') is indicative of (closet) dualism—and seeing that I have rejected dualism as unsatisfactory—I am not able to endorse Kihlstrom's remarks about whether or not 'consciousness' *accompanies* perceptual-cognitive functions (any more than whether 'pairhood' *accompanies* left-hand-plus-right-hand gloves); but what we could easily say is something like 'What-it-is-like/experience is not to be identified with 'consciousness' of any particular (sensory) perception, since perceptual functions can take place completely in the absence of any associated 'conscious' what-it-is-like/experience. Rather, whether or not any experiential quality attaches to such functions depends on whether or not *attention* is being paid to these functions.'

Whereas Chalmers said that questions about being 'conscious' of perceptions, cognitions, et cetera were only 'the easy problems' of consciousness, I have now shown that, insofar as these easy problems can better be understood in terms of *attention*, they may not be problems of 'consciousness' *at all* (this seems to me to be a particularly plausible suggestion, in light of my claim that the vicious reification of 'consciousness' or 'the mind' implies the endorsement of some minimum level of dualism, Sect. 2.2.3). To the extent that understanding these perceptual-cognitive phenomena in terms of attention is more satisfactory than understanding them in terms of 'consciousness,' understanding them in terms of 'consciousness' is problematic.

Smithies (2011: 31) writes that "Treisman's theory suggests that attention to an object, or its location, is necessary for trading on the identity of the object because it is necessary for solving the *binding problem*—that is, the problem of binding properties together as properties of a single object"—but can I just point out that *only a closet dualist* could make sense of the idea that 'properties' could in any way be 'bound' together 'as properties of a single object'? For any genuine physicalist, for whom there is no such 'thing' as 'a property' [compare: Sect. 2.2.1.3], there can obviously also be no such thing as 'the binding problem.' Anybody who believes that 'a property' could even potentially be 'bound' together with another 'property' cannot plausibly claim to believe that everything that exists is physical.

3.3.3 The Hard Problem of What-It-Is-Like/Experience: Conclusion

In addition to showing the problems inherent in trying to account for '*consciousness*' in perceptual or cognitive terms [Sect. 3.2], I have now shown that it is similarly problematic to try and account for *what-it-is-like/experience* in such terms [Sect. 3.3]. Since sensory perception, and cognition, seem to be almost completely independent of 'consciousness' [Sect. 3.2]—and since there is *nothing* it is like to perceive things, or perform cognitive operations, 'unconsciously'—attempting to account for (hard-problem) 'consciousness' in terms of perception or cognition is problematic.

Talking about whether or not *attention* is being paid to a certain perceptual/cognitive operation (rather than whether or not it is 'conscious') seems to be a more satisfactory way of accounting for these (easy-problem) phenomena. On a conceptual level as well, it is certainly preferable to talk about whether somebody is *paying attention* to a certain perception/cognition rather than whether or not 'consciousness' is *accompanying* that perception/cognition.

Because perception and cognition seem to be almost entirely independent of 'consciousness'—and because there is nothing-it-is-like to experience unconscious (and/or unattended) perceptual and/or cognitive processes—we should be careful not to conflate (hard-problem) what-it-is-like/experience with (perceptual-cognitive) 'consciousness.'

3.4 The Hard and the Easy Problems of Consciousness: Conclusion

In Chap. 2, I showed that, so long as we are not dualists, we should not accept that there is an 'explanatory gap' between 'the mind' and 'the body': to talk about 'the mind' as if it were a discrete causal entity is to commit the fallacy of reification, and committing the fallacy of reification with regard to 'the mind' amounts to the same thing as embracing dualism. These same considerations apply to 'consciousness' and 'the brain: ' strictly speaking, nobody with naturalist convictions should be able to talk as if 'consciousness' and 'the brain' are both legitimate entities. But if we overlook this reification temporarily, and examine some typical 'consciousness' research, we find that:

1. Because sensory perception and cognition seem to be almost entirely independent of 'consciousness,' we should be careful not to conflate *consciousness* with ('conscious') *perception* or *cognition* [Sect. 3.2]; and
2. Because sensory perception and cognition seem to be almost entirely independent of 'consciousness'—and because there is nothing-it-is-like to experience unconscious (and/or unattended) perceptual and/or cognitive processes—we should be careful not to conflate (hard-problem) *what-it-is-like/experience* with (perceptual-cognitive) '*consciousness*' [Sect. 3.3].

Therefore, since we are trying to find out about the hard problem of what-it-is-like/experience, we would do well not to devote too much energy to investigating anything to do with sensory perception and/or cognition. If we truly want to make any headway against Chalmers' 'hard problem,' we need to abandon these kinds of investigations. Just as I abandoned my investigation of 'the problem of the explanatory gap,' upon discovering that only dualists could coherently recognize such a gap (in Chap. 2), I also need to abandon the investigation of perceptual-cognitive (/sensory/easy-problem) 'consciousness'—since I have now shown that so much perception and cognition takes place without conscious awareness.

Accordingly, in the next chapter, I will explore Freud's view about how strongly (psychological) what-it-is-like/experience is dissociated from (introspective) 'consciousness.'

Notes

1. Montemayor and Haladjian agree that "most contemporary analyses of consciousness … focus mainly on perception" (2015: ix).

2. In English, "saying ... that the object of perception is an idea is not as well adapted to evoking chuckles" as it is in certain African languages (Wiredu 2011: 32–33)—which is to say, Wiredu does not approve of any view according to which 'consciousness' of ideas is seen as being similar to 'consciousness' of perceptions.
3. Note that Block, like Chalmers, appears to be using 'consciousness' and 'conscious visual perception' interchangeably [compare: Sect. 3.1].
4. See also Weiskrantz (2002: 571), Phillips (2018: 482).
5. The influence of subliminal perceptions on behavior is generally considered to be reasonably well-established. Newell and Shanks caution that such studies may not provide *robust evidence* of unconscious influences on behavior, but they do not appear do be suggesting that *there are no such influences*.
6. *Affective*: relating to, arising from, or influencing feelings or emotions.
7. Again, note that I am not attempting to assess Churchland's overall view, but merely highlighting how this specific statement she has made exhibits that bias—which Chalmers mentioned, and which I have identified in a number of authors' work—towards understanding consciousness awareness (and/or what-it-is-like/experience) in terms of 'easy-problem consciousness.'

References

Bargh, J.A. 2014. Our unconscious mind. *Scientific American* 310: 30–37. https://doi.org/10.1038/scientificamerican0114-30.

Bargh, J.A., and T.L. Chartrand. 1999. The unbearable automaticity of being. *American Psychologist* 34 (7): 462–479.

Bargh, J.A., Chen, M., and Burrows, L. 1996. Automaticity of social behavior: Direct effects of trait construct and stereotype activation on action. *Journal of Personality and Social Psychology* 71: 230–244.

Block, N. 1995. On a confusion about a function of consciousness. *Behavioral and Brain Sciences* 18: 227–287.

———. 2009. Comparing the major theories of consciousness. In *The cognitive neurosciences IV*, ed. M. Gazzaniga, 1111–1122. Cambridge: MIT Press.

Chalmers, D.J. 1995. Facing up to the problem of consciousness. http://consc.net/papers/facing.pdf. Accessed 23 March 2015. [Also published in *Journal of Consciousness Studies* 2 (3): 200–219].

———. 1996. *The conscious mind: in search of a fundamental theory*. https://pdfs.semanticscholar.org/5491/3945b27ca973968174ad521aa85620f0b32f.pdf. Accessed 7 August 2016.

Cherry, E.C. 1953. Some experiments on the recognition of speech with one and two ears. *Journal of the Acoustical Society of America* 25: 975–979.

Churchland, P.S. 1986. *Neurophilosophy: Toward a unified science of the mind-brain*. Cambridge: MIT Press.

Cowey, A. 2010. The blindsight saga. *Experimental Brain Research* 200: 3–24. https://doi.org/10.1007/s00221-009-1914-2.

Cowey, A., and P. Stoerig. 1995. Blindsight in monkeys. *Nature* 373: 247–249. https://doi.org/10.1007/s00221-009-1914-2.

De Gelder, B., M. Tamietto, G. van Boxtel, R. Goebel, A. Sahraie, J. van den Stock, B.M.C. Stienen, L. Wieskrantz, and A. Pegna. 2008. Intact navigation skills after bilateral loss of striate cortex. *Current Biology* 18 (24): R1128–R1129. https://doi.org/10.1016/j.cub.2008.11.002.

Dijksterhuis, A., et al. 2006. On making the right choice: The deliberation-without-attention effect. *Science* 311: 1005–1007. http://www.sciencemag.org.

Dresp-Langley, B. 2012. Why the brain knows more than we do: nonconscious representations and their role in the construction of conscious experience. *Brain Sciences*, 2: 1–21. http://www.mdpi.com/2076-3425/2/1/1.

Gallagher, S., and D. Zahavi. 2008. *The phenomenological mind: An introduction to philosophy of mind and cognitive science*. Routledge. https://doi.org/10.4324/9780203086599.

Gazzaniga, M.S., R.B. Ivry, and G.R. Mangun, eds. 2002. *Cognitive neuroscience: The biology of the mind*. 2nd ed. New York: W.W. Norton.

Gennaro, R.J. 2004. Higher-order theories of consciousness: An overview. http://www.usi.edu/libarts/phil/gennaro/papers/johnbenjintro.pdf. Accessed 6 December 2016.

Kahneman, D., and Treisman, A. 1984. Changing Views of Attention and Automaticity. *In Varieties of attention*. Edited by Parasuraman, R. & Davies, D.R. http://faculty.washington.edu/jpalmer/files/forCM/FilteringPaperNewRefs/kahneman_treisman1984.pdf.

Kihlstrom, J.F. 1987. The cognitive unconscious. *Science* 237: 1445–1452.

Koch, Christof, and Naotsugu Tsuchiya, 2007. Attention and consciousness: Two distinct brain processes. *Trends in Cognitive Sciences* 11(1): 16–22.

Kouider, S., and S. Dehaene. 2007. Levels of processing during non-conscious perception: A critical review of visual masking. *Philosophical Transactions of the Royal Society* 362: 857–875. https://doi.org/10.1098/rstb.2007.2093.

Marcel, A.J. 1983. Conscious and unconscious perception: Experiments on visual masking and word recognition. *Cognitive Psychology* 15: 197–237.

Merker, B. 2005. The liability of mobility: A selection pressure for the transition to consciousness in animal evolution. *Consciousness and Cognition* 14: 89–114. https://doi.org/10.1016/S1053-8100(03)00002-3.

Montemayor, C., and H.H. Haladjian. 2015. *Consciousness, attention, and conscious attention*. ProQuest Ebook Central Database. https://0-ebookcentral-proquest-com.ujlink.uj.ac.za/lib/ujlink-ebooks/reader.action?docID=3339975.

Newell, B.R., and D.R. Shanks. 2014. Unconscious influences on decision making: A critical review. *Behavioral and Brain Sciences* 37: 1–19. https://doi.org/10.1017/S0140525X12003214.

Nisbett, R.E., and T.D. Wilson. 1977. Telling more than we can know: Verbal reports on mental processes. *Psychological Review* 84 (3): 231–259.

Phillips, I. 2018. Unconscious perception reconsidered. *Analytical Philosophy* 59 (4): 471–514.

Prinz, Jesse J. 2012. *The Conscious Brain: How Attention Engenders Experience*. New York: Oxford University Press.

Ryle, G. 1949. *The concept of mind*. London: Routledge.

Schvaneveldt, R.W., and D.E. Meyer. 1973. Retrieval and comparison processes in semantic memory. In *Attention and performance IV*, ed. S. Kornblum, 395–409. New York: Academic.

Searle, J.R. 1997. *The mystery of consciousness*. 7th ed. New York: NYREV.

Smithies, D. 2011. What is the role of consciousness in demonstrative thought? *Journal of Philosophy* 108 (1): 5–34. https://doi.org/10.5840/jphil201110811.

Sperling, G. 1960. The information available in brief visual presentations. *Psychological Monographs: General and Applied* 74: 1–29.

Tallon-Baudry, Catherine. 2012. On the neural mechanisms subserving consciousness and attention. *Frontiers in Psychology* 2: 397.

Tanney, J. 1949. Rethinking Ryle: A critical discussion of *The Concept of Mind*. In *The concept of mind*, ed. G. Ryle and ix-lvii. London: Routledge.

Weiskrantz, L. 2002. Prime-Sight and Blindsight. *Consciousness and Cognition*, 11 (4): 568–581. https://doi.org/10.1016/S1053-8100(02)00015-6.

Westen, D. 1998. The scientific legacy of Sigmund Freud: Toward a psychodynamically informed psychological science. *Psychological Bulletin* 124 (3): 333–371.

Wiredu, K. 1987. The concept of mind with particular reference to the language and thought of the Akans. In *Contemporary philosophy: A new survey*, ed. G. Fløistad, vol. 5, 153–179. Dordrecht: Martinus Nijhoff Publishers.

———. 2011. Empiricalism: The empirical character of an African philosophy. In *Identity meets nationality: Voices from the humanities*, ed. H. Lauer, N.A.A. Amfo, and J.A. Anderson, 18–34. Legon-Accra: Sub-Saharan Publishers.

4

(Un)Consciousness and (Ir)Rationality in Psychology

I suggested [Sect. 2.2] that even recognizing the existence of 'the mind' and/or 'consciousness' constituted a tacit endorsement of dualism, which I would not be considering as a plausible position. That seems like a good conceptual reason for not conflating what-it-is-like/experience either with 'the mind' or with 'consciousness.'

But even if we continue to refer to 'consciousness,' for the sake of argument, I said (Chap. 3), that investigators in the mind sciences often think of 'consciousness' in terms of being 'conscious' of various sensory perceptions, or in terms of 'consciously' performing cognitive operations. I said that perception and cognition can be, and often are, *un*conscious. Moreover, when perception and cognition take place 'unconsciously,' there is *nothing*-it-is-like to perceive those things, or perform those cognitive operations [compare: Sect. 3.3]—which seems to be a good empirical reason for being careful not to conflate what-it-is-like/experience with perceptual-cognitive 'consciousness.' In this chapter, I will examine some psychological considerations which appear to indicate that we should also be careful not to conflate *(psychological) what-it-is-like/experience* with (introspective) '*consciousness*' [Sect. 4.2].

Mind science investigators also often understand 'consciousness' as synonymous with *rationality*. For example, Churchland claims that the

© The Author(s), under exclusive license to Springer Nature Switzerland AG 2022

J. Anderson, *Biological Naturalism and the Mind-Body Problem*,
https://doi.org/10.1007/978-3-030-99684-0_4

main motivation for accepting dualism is that it is so difficult to account for "reasoning and consciousness and their kind" (1984: 319) in the same naturalistic terms we use to account for physical matter. Evidently, she takes it for granted that 'consciousness' and 'reasoning' are, if not actually the same thing, at least very similar. The same can be said for many other philosophers: they think of 'consciousness' as quintessentially some kind of process during which reasoning is correctly applied to premises in order to reach correct conclusions.

The assumption that people do things in accordance with the principles of reason is a prerequisite for taking any kind of 'computational' view—which quite a few philosophers still do [see Sects. 2.3.2, 2.3.3, and 2.3.4]. When anybody is trying to produce any kind of computational account of consciousness, that person must necessarily be assuming that 'conscious minds' work 'rationally' (according to the rules of logic), just as computers do. But this is problematic—and in Sect. 4.3, I will explain why.

4.1 Introduction: Radical Behaviorism

When the word 'psychology' began to be used to refer to a specific branch of science (Ryle puts this at about the middle of the eighteenth century), the idea was that psychologists would study 'mental phenomena' in the same way that other types of scientists studied physical phenomena.

> As Newtonian scientists studied the phenomena of the one field, so there ought to be scientists studying the phenomena of the other field. Moreover, as Newtonian scientists found and examined their data in visual, auditory and tactual perceptions, so psychologists would find and examine their counterpart data by counterpart, non-visual, non-auditory, non-tactual perception. (Ryle 1949: 293)

But admirable as these intentions may have been, those experimental psychologists had no choice but to abandon this program (of studying non-visual, non-auditory, non-tactual perception), because "[a] researcher's day cannot be satisfactorily occupied in observing nonentities and describing the mythical" (Ryle 1949: 294). This is a problem which still looms large in psychology today: it is supposed to be a scientific discipline, but as such, it has no proper object of study. The 'science' of 'mental phenomena' "is a discipline with no methods, no data, no results, no future, no promise" (Dennett 2001, in Gallagher and Zahavi 2008: 14).

What the discipline of psychology desperately needed was a new conceptual framework within which to work towards a scientific account of psychological phenomena. So when a small group of psychologists (especially in America) started to come up with the idea of behaviorism, it was seized upon very enthusiastically by psychologists. After 1913 (which is when John Watson's highly influential *Psychology as the Behaviorist Views it* was published), experimental psychology saw a shift in emphasis, away from introspective consciousness, and towards the objective observation of behavior.

According to the (radical) behaviorist, the only proper purpose of psychology is the prediction and control of behavior. During the height of the behaviorist era, many academic psychologists discouraged (and even forbade) the investigation of such 'pseudoscientific' issues as 'consciousness'; the behaviorist movement therefore "established the idea that psychological explanation can proceed while the phenomenal is ignored" (Chalmers 1996: 12). In effect, this is much the same as saying that, precisely because 'the hard problem' of consciousness is so hard, almost all so-called 'consciousness' research actually turns out to be research into one of 'the easy problems' only [compare: Sect. 3.1]. The hard problem of consciousness is so hard to investigate that experimental psychology, in effect, just *does not* investigate it.

Behaviorist views "captivated mainstream American psychology for many years" (Panksepp 1998: 10); and currently, although the popularity of Behaviorism has diminished quite a lot since its heyday, the behaviorist paradigm is still very firmly entrenched in twenty-first-century psychology. However, although the "behaviorist juggernaut" (Solms & Panksepp 2012: 170) may have dominated the field of psychology comprehensively for a significant part of the twentieth century, the behaviorist juggernaut arguably did not have as much influence over the field as did the previous juggernaut in the discipline—and the previous juggernaut in psychology was, inevitably, Sigmund Freud. Whereas the 'psychological' theories produced by the behaviorists completely disregarded (psychological) what-it-is-like/experience, (psychological) what-it-is-like/experience is precisely what Freud's theories were aimed at accounting for.

In this chapter, I am going to examine Freudian ('Psychodynamic') Theory. The structure of this chapter differs from the structure of the previous two chapters in that, instead of mentioning a number of studies/paradigms/points of view, I am going to focus in great detail on a single theory. Because Freudian psychodynamics is so incredibly rich and detailed, I think I can find all the illustrations I need to make my arguments within this single body of work. This theoretical framework, I argue, is well-suited to conceptualizing precisely the type of (psychological) what-it-is-like/experience—as opposed to un-/conscious perception or cognition (that is to say 'the easy problems' of consciousness, Sect. 3.2), *and* as opposed to the 'observable behavior' which was all the behaviorists cared about—that I am interested in. Within the psychodynamic framework, it is possible to account for (psychological) what-it-is-like/experience without having recourse to (un-/conscious) perception/cognition, which is a point in favour of this theory, seeing that, as I demonstrated [Sect. 3.3], it is not a good idea to conflate what-it-is-like/experience with (perceptual-cognitive) 'consciousness.'

In fact, as I will now explain [Sect. 4.2], psychodynamic theory is able to account for what-it-is-like/experience without having recourse to (introspective) 'consciousness' of any description at all—which is a further point in favour of this theory. Not only have I already demonstrated that 'consciousness' does not 'exist' (for those who are not dualists, see Sect. 2.2.3); I have also demonstrated that 'consciousness' (in its scare-quotes sense) and 'what-it-is-like/experience' can be strongly dissociated [Sect. 3.3.2]. Moreover, within the psychodynamic framework, it is possible to account for what-it-is-like/experience without having recourse to the concept of rationality [Sect. 4.3]. This is also a point in favour of this theory, since (as I will explain) it is not very plausible to think of human beings as intrinsically rational. This being the case: I will argue that we should be careful not to conflate (psychological) what-it-is-like/experience—in Nagel's sense, or Chalmers' 'hard-problem' sense—either with (introspective) 'consciousness' or with 'rationality.'

4.2 (Psychological) What-it-is-like/experience and (Introspective) Consciousness

According to Freud, "[i]t is generally taken for granted that 'the psychical' involves "the quality of *being conscious*—unique, indescribable, but needing no description" (1940b, in *Complete Works*, p. 5068). Moreover, "[t]o most people who have had a philosophical education," he claims, "the idea of anything mental which is not also [introspectively] conscious is so inconceivable that it seems to them absurd and refutable simply by logic" (1923b, in *Complete Works*, p. 3947). Everything 'psychical,' they assume, must also be 'conscious'; and everything 'conscious' must also be 'psychical.' It seems reasonable to suggest that this bias towards equating 'mental' with 'conscious' is still with us, given the extent to which philosophers tend not to pay much attention to the kind of 'unconscious' psychological phenomena Freud was interested in. But equating 'psychical' with 'conscious' is not much help as far as regards understanding what 'the psychical' actually *is*.

In order to shed any light on 'the psychical,' says Freud, "the philosophers at least were obliged to assume that there were organic processes parallel to the conscious psychical ones, related to them in a manner that was hard to explain, which acted as intermediaries in the reciprocal relations between 'body and mind'." However, Freud quite rightly dismissed this approach as "unsatisfactory" (1940b, in *Complete Works*, p. 5068), because such an understanding requires a level of dualism which is not acceptable to any naturalist. On the Freudian account, in contrast, rather than being taken as the defining characteristic of all psychological phenomena, 'consciousness' is understood as being "only a *quality* of what is psychical, and an inconstant quality at that—one that is far oftener absent than present." By denying unequivocally that 'consciousness' is the essence of the psychical, Freudian theory is able to maintain that psychical processes are "probably similar in kind to all other natural processes of which we have obtained knowledge" (ibid.).

In this section, I will first [Sect. 4.2.1] examine what Freud himself explicitly said about 'what is conscious'—and investigate his reasons for claiming that psychodynamic theory cannot situate the essence of 'the psychical' in what is conscious.[1] Secondly, by looking the central techniques in Freudian psycho-analysis [Sect. 4.2.2], I will demonstrate the kind of considerations that led Freud to develop his theory of 'the mental apparatus'; as well as the way in which he is able to account for the symptoms of some actual patients most satisfactorily in terms of this theory.

4.2.1 Sigmund Freud

Owing to the fact that Freud's ideas have made such an impact on so many aspects of society, there is a truly staggering amount of ill-informed criticism of him; and those who do think that Freud had some good ideas find themselves spending a disproportionate amount of their time occupied with "the somewhat negative task of defending Freudian theory" (Levine 2000: 2) against those who are intent on discrediting it. Freud himself, quite early in his career, already had occasion to note that "[t]here is often a tendency in medical and especially in psychiatric circles to contradict the theories of psycho-analysis without any real study or practical application of them" (observing, with his usual astuteness, that this might easily be due to "the dread felt by the average observer of seeing himself in his own mirror," 1913, p. 2548).

Much of the huge body of criticisms of Freud and his views focus on specific elements of his clinical theories/practices (for example, whether the 'oral' stage of infant development can be said to last from birth to one year of age, or rather the first 18 months)—none of which is relevant to the aspects of psychodynamic theory which interest me, which is why I will not discuss such criticisms. In Sects. 4.3.2 and 5.2.2 I will discuss in great detail some specific challenges to Freudian theory from Mark Solms which are much more pertinent to this discussion.

On the other hand, a number of thinkers who have studied and practically applied the theories of psycho-analysis agree that some of Freud's ideas were truly visionary. Ryle calls Freud "psychology's one man of genius," and the influence of his teachings "deservedly profound" (1949: 297); and there are many who would agree with Eric Kandel that "psychoanalysis still represents the most coherent and intellectually satisfying view of the mind" (1999: 505) we have even today. Antonio Damasio (whose work forms a major part of Chap. 6) writes that some of Freud's insights "are consonant with the most advanced contemporary neuroscience views" (Damasio 1999, p. 38, in Solms 2013: 9); and Jaak Panksepp, (with whose views my own views seem to be the most closely aligned, see Sect. 6.3), says that Freud's work "revolutionized our views of how the mind operates" (2004: 8).

What Freud is mostly famous for is his practice of 'psychoanalysis'—but in fact psychology was not his first choice of career. When he enrolled at the University of Vienna in 1873, he planned to study law, but the allure of a medical career proved too much for him to resist. He registered as a medical student, intending "to embark not on a conventional career as a physician, but on philosophical-scientific investigations that might solve some of the great riddles that fascinated him" (Gay 1989: x)—including, presumably, the 'great riddle' of the explanatory gap between mind and body, and/or 'the mystery' of consciousness.

From 1876–1882, Freud worked at the Institute of Ernest Brücke, who was a professor of physiology at the University of Vienna (Taylor 2009: 53). During his time there, he published influential scientific papers, on subjects ranging from the neuroanatomy of *Ammocoetes* (a primitive fish) to a new chemical method for preparing nerve tissues for microscopic examination; and it was only with reluctance that Freud grudgingly acknowledged the inevitable financial difficulties attendant on a career in biological research, and turned instead to medicine. However, he never stopped being keenly interested in biology. Even in the midst of his success and growing stardom in his clinical practice, Freud remarked, in conversation with a friend, "[I hope] to return to my old pursuit and do a little anatomy; after all, that is the only satisfying thing" (in Sulloway 1979: 16).

Because of the wide range of his interests, Freud's work has important implications for psychologists, neuroscientists, and animal science researchers—but it is also important to philosophers. Levine (2000: 1) claims that "Freud was as much a philosopher as he was a psychologist"—if not more so. That is why he finds it surprising (as do I) that "relatively little" has been written, *by philosophers*, about the philosophical implications of the Freudian framework. If the core tenets of psychodynamic theory are, as I believe (and Levine agrees), essentially correct, then "studies of its wider philosophical implications—studies prevented in various ways by arguably irrelevant criticism—are worthwhile and overdue" (Levine 2000: 2).

Terminological Note:

Because Freudian psychoanalysis has had such a widespread impact on laypersons, as well as professionals, many people use the term 'psychoanalysis' indiscriminately in the sense of 'anything to do with Freud and his theories.' Before I start discussing Freud's work, therefore, an important terminological note:

Freud said that

> Psycho-analysis is the name (1) of a procedure for the investigation of mental processes which are almost inaccessible in any other way, (2) of a method (based upon that investigation) for the treatment of neurotic disorders and (3) of a collection of psychological information obtained along those lines, which is gradually being accumulated into a new scientific discipline. (1923a, in *Complete Works*, p. 3912)

There are plenty of writers who use 'psychoanalysis' in a similar way; for example, Levine comments that

> Psychoanalysis refers (i) to a theory of the structure and workings of the mind, and (ii) to the psychotherapeutic method—both originally based on Freud's theoretical and practical (clinical) work. (Levine 2000: 2)

(Here Levine seems to have made Freud's points (1) and (2) into a single point, Levine's point (ii).)

What Freud himself said about psycho*dynamic* theory was "[t]he view is a dynamic one, in so far as it regards psychical processes as displacements of psychical energy which can be gauged by the amount of their effect on the affective elements" (1913, in *Complete Works*, p. 2545). In what follows, I will use 'psycho*analysis*' to refer to the actual, practical 'analysis' of a patient, using accepted psychoanalytic techniques. These psychoanalytic techniques (i) are derived from what I will refer to as 'psycho*dynamic*' theory,[2] and (ii) are themselves used to inform the construction of psychodynamic theory.

According to Jim Hopkins, psychodynamic theory is supported by a strong explanatory framework, and is well positioned "to play a significant role in the understanding of the working of motive in individual psychology" (2015: 62).

4.2.1.1 An Outline of Psycho-Analysis

An Outline of Psycho-analysis (1940a) was written towards the end of Freud's career, and therefore contains some of his most well-considered ideas. In this work, Freud aims to "bring together the tenets of psycho-analysis … in the most concise form and in the most unequivocal terms." The teachings of psycho-analysis are, he claims, based on "an incalculable number of observations and experiences" (1940a, in *Complete Works*, p. 4956), which is why there is a very real sense in which it is impossible for anybody to explain psychodynamics thoroughly in a single written work. In *Outline*, Freud is not even attempting a thorough explanation: he explicitly stipulates that he is omitting all unnecessary examples and 'proofs,' in the name of being concise. But it is a good place to look if we want to know which principles are the most central in psycho-analysis.

"We know two kinds of things about what we call our psyche (or mental life)" (ibid.), Freud says: firstly, we know that the brain (or nervous system) is the bodily organ of mental life. "We assume that mental life is the function of an apparatus to which we ascribe the characteristics of being extended in space, and being made up of several portions" (1940a, in *Complete Works*, p. 4957). Just as (I claimed, contra Cottingham, Sect. 2.2.1.1) Descartes knew that there was a close connection between the thinking thing and the physical brain, so Freud's whole view of psychology rested on the biological character of (human) organisms. Freud not only did not even entertain the idea that 'the mind' was different in principle from 'the body,' but explicitly situated his account within what I will refer to (in Chap. 5) as 'biological naturalism.'

The second thing we know about the psyche (or 'mental life') is "our acts of consciousness" (Freud 1940a, in *Complete Works*, p. 4956). Our knowledge of 'consciousness' is direct and immediate, and cannot be explained by reference to any further facts. In a way which is very reminiscent of Nagel's (not to mention, of many contemporary philosophers') formulation of the explanatory gap problem [Sect. 2.1], Freud says that "[t]he starting point for this investigation is provided by a fact without parallel, which defies all explanation or description—the fact of consciousness." In spite of the fact that it is so difficult to explain or describe, "if anyone speaks of consciousness we know immediately and from our most personal experience what is meant by it" (1940a, in *Complete Works*, p. 4967).

But although we view 'consciousness' as something unique, special, or different in kind from other things, we also generally accept that 'conscious processes' cannot be completely independent of physical processes. Freud thinks that any reasonable person must agree that "there are physical or somatic processes which are concomitant with the psychical ones" (ibid.). Moreover, a physical account of psychical phenomena should be viewed as "more complete" than the corresponding psychical account—because, although it is sometimes possible to discover physical processes that do not have any corresponding psychical processes, there is no such thing as a psychical process which does not correspond to any physical process. Therefore, as Freud observes, "it … becomes plausible to lay the stress in psychology on these somatic processes [and] to see in *them* the true essence of what is psychical" (ibid.).

In other words, he recognizes that it is not acceptable to propose that conscious processes are different in principle from physical processes. This amounts to no more than saying that Freud is not a dualist—and that, therefore, he cannot recognize any kind of explanatory gap. Just as, for the purposes of this book, I am taking it for granted that dualism is not a plausible metaphysical position [Sect. 2.2.1.4], so Freud seems to have taken it for granted that a satisfactory account of 'the mental apparatus' needed to be a naturalistic one (see also Sect. 4.2).

For Freud, 'the psychical' predominantly comprises unconscious processes, which are similar in kind to other natural/physical processes—a minority of which enter 'consciousness,' but still retain their essential natural, physical nature. Accordingly, it is difficult to see what Boag has in mind when he accuses Freud of "propos[ing] a *dualism* between conscious and unconscious processes that cannot be coherently maintained" (2015: 241, my emphasis). Freud's commitment to explaining psychological phenomena in robustly physicalist terms could hardly have been stronger than it is, as I will continue to emphasize.

Similarly, Wakefield (2018: 8) thinks that "Freud explicitly leaves aside the challenge that has absorbed the attention of much of the field of philosophy of mind for the last half of the twentieth century of specifying what property confers intentional content on a brain state"—but insofar as talking about the 'intentional content' of 'brain states' constitutes vicious reification (amounting to the endorsement of dualism), and insofar as Freud is not a dualist, Wakefield's criticism fails to hit home. It makes no sense to claim that Freud "must explain the distinction between those nonconscious brain states that influence consciousness but are themselves nonmental and those nonconscious brain states that influence consciousness and are genuinely mental" (Wakefield 2018: 217). Freud would simply deny that 'nonconscious brain states' or 'consciousness' even *existed*—which implies that he would, *a fortiori*, deny that any 'brain state' was the kind of thing that could correctly be described as either 'mental' or 'nonmental'—and remind Wakefield that 'the psychical' was probably similar in kind to all other natural (physical) processes [Sect. 4.2].

4.2.1.2 Science and Psychoanalysis

Although any number of specific criticisms have been levelled against psychoanalysis, "a central claim has long been that … psychoanalytic constructs … lack any basis in scientific research" (Westen 1999: 1061). But to complain that Freudian theory is insufficiently 'scientific' is to miss the point of the theory completely. It is legitimate to characterize certain kinds of institutions and movements as 'scientific,' but for others, "the views of science are, for the most part, simply irrelevant" (Talvitie 2015: 186). For instance, sports clubs or music groups have nothing to do with science; therefore, it makes no sense to claim that these organizations are non-scientific. On this understanding, psycho*analysis* (as a therapeutic technique) falls into the same category as sports clubs and music groups: *not* the kind of thing which could potentially be described as 'scientific.' On the other hand, psycho*dynamic* theory (as a theory about the somatic processes which should be seen as the true essence of the psychical) *is* the kind of thing which could be described as 'scientific'—and, as I explained [Sect. 4.2.1], psychodynamic theory was securely founded on Freud's ('scientific') biological expertise.

Psychological processes, said Freud, are no more (but also no less) 'unknowable' than the processes which constitute any other kind of science. In physics or chemistry, for example, people can deduce what micro-level processes must be occurring (for example, some of the behavior of electrons), even if they can never directly observe these processes. They observe what is observable, and from that, they deduce what else must be taking place. Likewise, it may not be possible to observe psychological processes directly, but Freud maintains that "it is possible to establish the laws which they obey and to follow their mutual relations and interdependences unbroken over long stretches" (1940a, in *Complete Works*, p. 4968). Physical and chemical terms are also 'figurative,' strictly speaking, but at least the figurative language of physiology/chemistry is familiar—and probably simpler, as well. Perhaps that explains why it is easy to fall into the trap of objecting to psychoanalysis's use of figurative language, while ignoring the fact that all scientific disciplines essentially do the same thing.

Whereas Descartes thought it quite impossible to deny that thinking/consciousness was different in principle from physical objects, Freud, even in 1940, thought that dualism was an untenable philosophical position. He put 'body and mind' in scare quotes [Sect. 4.2], just as I do—knowing, as he did, that there is a very real sense in which 'the mind' does not 'exist' (independently from the physical body); and he thought that "the deficiencies in our description [of psychical phenomena] would probably vanish if we were already in a position to replace the psychological terms by physiological or chemical ones" (1920, in *Complete Works*, p. 3760). However, as he quite rightly pointed out, when one is not yet in a position to replace psychological terms with physiological/chemical ones, one has no choice but to make do with a somewhat deficient description. Without using "the figurative language, peculiar to … depth psychology" (1920, in *Complete Works*, p. 3760), it would not be possible to produce psychological theories *at all.*

What Freud was really interested in was understanding in depth people's subjective, psychological, what-it-is-like/*experience*; and in order to do this, he found it necessary to postulate some hypothetical forces and processes. But whereas the behaviorists, in their attempt to be scientific, had little choice but to *ignore* any psychological processes which couldn't be observed in an 'objective' way, Freud used his expertise in biology to help him develop a plausible (naturalistic) *theoretical* account of such processes. And whereas many philosophers of mind assume that there is no place for what-it-is-like/experience in a purely physical world, Freud knew that (psychological) what-it-is-like/experience exists precisely *because* everything that exists is physical. His view "enabled psychology to take its place as a natural science like any other" (1940a, in *Complete Works*, p. 4968), since he wanted, just as I do, to account for thinking/consciousness in terms of the functioning of physical structures [compare: Sect. 2.2].[3]

4.2.1.2.1 Freud and Darwin

> Psychoanalysis is one of those rare intellectual achievements that had the effect of radically transforming human self-understanding. … psychoanalysis effectively completed the intellectual revolution begun by Copernicus, and advanced by Darwin, a revolution that undermined human conceit regarding its putatively special and privileged position in the cosmos and in nature. (Lapsley and Stey 2011: 1)

Not only was Freud's influence comparable to Darwin's: psychodynamic theory is actually rather similar in form to the theory of evolution by natural selection. Karl Popper's criticism of psychodynamics, on the grounds that it is not 'scientific,' has been very influential; but actually, Popper said much the same about Darwinian evolutionary theory: neither Darwinian theory nor Freudian theory is what Popper calls 'scientific.' Rather, they are 'metaphysical' theories, because they are not 'falsifiable'; moreover, neither theory is capable of predicting the data they undertake to explain with a probability value that even approaches 1. But saying that neither Darwin's theory of evolution nor Freud's psychodynamic theory is 'scientific' need not imply that neither theory is sensible, or insightful, or that neither theory revolutionized human knowledge or understanding.

According to Hopkins, Darwin's account of why we should accept his hypotheses was actually far better than Popper's. Darwin said that we should accept his hypotheses not because they were so 'scientific,' but "because they explained a great range of phenomena that no others did" (2015: 57). Moreover, Darwin thought that we should accept his theory because, although the theory did not make predictions whose value even approach a value of 1, the theory was predictive in a "weak but essential sense": in much the same way that Newton and Einstein started out with hypotheses about celestial objects which moved as they did because they were (somehow) attached to (some kind of) spheres that (for some reason) circled the earth, so Darwin proposed that animals must have the characteristics they have because those characteristics (in some way) enhanced (some kind of) capacity that (for some reason) enabled them to survive better than their competitors. Freud, likewise, proposed various psychical forces and structures that in some way (he did not attempt to specify exactly how), served the psychological and biological needs of the psychodynamic subject.

Hopkins suggests that, from the very beginning, Darwin's theories were admired by those who had studied them carefully, along with the data from which those theories had been derived, in virtue of their explanatory power—in much the same way that, according to Freud, the teachings of psychoanalysis could only be fully appreciated by those who were familiar with the observations and experiences from which they were derived.

4.2.1.2.2 The Pleasure Principle

"*Beyond the Pleasure Principle* [1919–1920] may be regarded as introducing the final phase of Freud's views" about 'metapsychology'; and in it one begins to "see signs of the new picture of the anatomical structure of the mind which was to dominate all Freud's later writings" (Freud 1920: 5, Editor's note). 'The pleasure principle' is the apparently uncontroversial assumption that organisms are driven to do things only if their actions bring them some 'benefit,' in the broadest sense: some kind of pleasure or other satisfaction must result from the action, otherwise the organism would have no reason to perform it. When someone chooses one course of action, rather than another, this is not only *because* the course of action in question is the one which the person believes will result in the highest pleasure: it is also *conclusive evidence* that this is what she believes. For Freud, the benefit or satisfaction humans derive from their actions always takes the same form: the stabilizing of an unpleasurable state of (very low or very high) 'intrapsychic tension.' Freud saw the pleasure principle as the ultimate motivation for all behavior, and his understanding of the principle confirms his commitment to accounting for psychological phenomena in biological terms.

As a first step towards accounting for the need to balance pleasure and unpleasure, Freud directs us to imagine "a living organism in its most simplified possible form as an undifferentiated vesicle of substance that is susceptible to stimulation" (1920, in *Complete Works*, p. 3730). In such an organism, the surface of the undifferentiated vesicle turned toward the external world would be the sole medium through which stimuli could be perceived. According to accepted evolutionary theory, one would expect that, to the extent that receiving accurate information about the environment has survival value, the outer layer of Freud's elementary vesicle should come to embody the most favourable possible qualities for the reception of this information.

Reception of stimuli would be important to enable the vesicle to determine the direction and nature of the external forces acting upon it; but it would not be necessary for the vesicle to handle the full impact of these forces. It would be sufficient for it to take in only a small quantity of a stimulus as a sample. In fact, not only would it be *unnecessary* for the vesicle to undergo the full force of external stimuli: it would be *undesirable*. Therefore, the vesicle's outer layer would not only function as an ideal receptor of stimuli, but also as a protective shield between the primitive vesicle of organic matter and the potential dangers of *too much* stimulation. Freud claims that "*[p]rotection against* stimuli is an almost more important function for the living organism than *reception of* stimuli" (1920, in *Complete Works*, p. 3731).[4]

But the case of the complex organisms is different from the case of the primitive vesicle. Complex organisms, in addition to dealing with stimulation from the external world, must be prepared to receive stimulation originating from within the organism itself. Since the 'internal' processes (and resulting 'stimulations') of the organism's nervous system are what enable the organism to be alive in the first place, there can by definition be no protective shield against them. An organism that was completely 'shielded' *from itself* would be a dead organism. But because no protective shield could be effective against 'internal' stimulations, "the excitations in the deeper layers [of the complex organism] extend into the system directly and in undiminished amount, in so far as certain of their characteristics give rise to feelings in the pleasure-unpleasure series." These internal stimulations (that is, pleasure/unpleasure) "predominate over all external stimuli" (Freud 1920, in *Complete Works*, p. 3731).

Because Freud continued to use the pleasure principle to help him understand the clinical phenomena he encountered in his patients, this is a good example of how Freud's background in neurobiology and commitment to physicalism continued to inform his metapsychological theories.

4.2.1.3 Motivating a Psychology of the Unconscious

As I said [Sect. 4.2], the division of the psychical into what is conscious and what is unconscious is the fundamental premise of psychodynamic theory. Predictably, this premise has been the target of much criticism; but I am unable to find fault with Freud's reasoning in this respect. He says that "an unconscious conception is one of which we are not aware, but the existence of which we are nevertheless ready to admit on account of other proofs or signs" (1912, in *Complete Works*, p. 2577)—which seems reasonable. His account of the proofs and signs which indicate such 'unconscious conceptions' also seems reasonable:

In the first place, it is *necessary* to assume the existence of the unconscious because "the data of consciousness have a very large number of gaps in them." In both healthy people and psychiatric patients, "psychical acts often occur which can be explained only by presupposing other acts, of which, nevertheless, consciousness affords no evidence" (1915, in *Complete Works*, p. 2991). This is something which should be clear not only to those who have sufficient experience in psychoanalysis, but to anybody at all.

In the second place, assuming the existence of the unconscious is perfectly *legitimate*, because assuming the existence of mental contents to which we do not have introspective access is essentially what we do every time we ascribe mental contents to anyone other than ourselves (1915, in *Complete Works*, p. 2993). And if we routinely ascribe mental contents to other people, without having introspective awareness of these contents, it is difficult to see plausible grounds for objecting to the ascription *to ourselves* of mental contents of which we are not introspectively aware. We might almost take the existence of unconscious conceptions as axiomatic, because "when all our latent memories are taken into consideration it becomes totally incomprehensible how the existence of the unconscious can be denied" (1915, in *Complete Works*, p. 2992). Everybody 'knows' many facts and 'has' many memories which are not simultaneously present in conscious awareness at any one time; to this extent, we might legitimately say that these are 'unconscious conceptions'.[5]

In other words: Freud provides a persuasive argument in favour of the assumption that 'the psychical' incorporates a substantial unconscious component (since making this assumption is both necessary and legitimate). As I will explain in a lot more detail in the rest of this section, assuming the existence of unconscious psychical material not only makes conceptual sense out of otherwise unrelated 'conscious acts,' but also "enables us to construct a successful procedure by which we can exert an effective influence upon the course of conscious processes" (ibid.)—namely, psychoanalysis.

4.2.1.3.1 The Structure of Consciousness

In psychodynamic theory, "'[b]eing conscious' is … a purely descriptive term," denoting the peculiarly "immediate and certain" (Freud 1923b, in *Complete Works*, p. 3948) nature of certain ideas. Only as one intentionally focuses attention on some idea is it *Conscious* in this sense; and usually, this state of 'consciousness' is very transitory. But under the right conditions, ideas that have already ceased to be conscious can easily become conscious again. Freud says that he refers to psychological material which is 'unconscious,' in this temporary way, "indifferently, sometimes as the *Pcs.* ['*Preconscious*'] and sometimes as the *Cs.* ['*Conscious*']" (1915, in *Complete Works*, p. 3011).

Although Freud himself prefers to refer to this kind of content as 'latent,' there is a sense in which it would be accurate to say that this (preconscious) idea was 'unconscious.' It is unconscious as a matter of fact, but not as a matter of principle. In a different situation, that very same 'unconscious' preconscious idea might become fully 'conscious': perhaps we could even say that whether or not we are consciously aware of such ideas just depends on whether or not we are *paying attention* to them [compare: Sect. 3.3.2]. Therefore, to worry about whether it would be accurate to say that latent/preconscious ideas were most properly to be described as 'conscious' or 'unconscious' "would lead to nothing more profitable than a verbal dispute" (Freud 1923b, in *Complete Works*, p. 3948).

However, psychoanalysis shows that some ideas can exert a powerful force over the conscious mind even if they never become conscious themselves. This kind of content is *Dynamically Unconscious*, and there is nothing 'indifferent' about how to refer to this dynamic unconscious. Freud asserts that unconscious ideas of this 'dynamic' type cannot become conscious, as latent/preconscious ones do, no matter how much attention [Sect. 3.3.2] or how many HOTs [Sect. 3.2.1.1] one may direct towards them—because "a certain force" (ibid.) within the mental apparatus works to keep these ideas outside introspective awareness.

The Freudian conception of consciousness is often visualized diagrammatically as an iceberg, with the (dynamic) Unconscious, submerged beneath the water, comprising the largest portion; the Preconscious, just below the water's surface, significantly smaller; and finally the part that is (introspectively) Conscious, just the tiny tip of the iceberg fully exposed. So central is the dynamic unconscious to all Freud's thinking that one would almost be justified in characterizing psychodynamic theory as a theory *about the dynamic unconscious.*

4.2.1.4 The Lesson (of 'Sigmund Freud') for Philosophy

As Freud quite correctly argued, postulating the existence of unconscious psychological phenomena is both necessary and legitimate. What was remarkable about Freud's understanding of 'the psychical' was his commitment to biological naturalism, in the midst of the firmly-entrenched mind-body dualism flourishing all around him. And key to this framework was his commitment to the claim that 'what is psychical' could not be equated with 'what is conscious.' His commitment to this claim is what should have enabled psycho-analysis to take its place among the natural sciences—although, as I have explained, this aspect of Freudian theory is seldom appreciated by critics. But several contemporary writers agree that Freud had "an interesting and sophisticated theory … of the 'what it is like' of consciousness"; and that his view of the unconscious was "grounded in interlacing theories of mental representation and consciousness that have a remarkably contemporary hue" (Redding 2000: 119).

4.2.2 Psychoanalytic Techniques

Seeing that, in the psychodynamic framework, the dynamic unconscious is the really important part of the mental apparatus, it is unsurprising to find that psychoanalytic techniques primarily function as ways of accessing dynamically unconscious psychological material. The three main psycho-analytic techniques are hypnosis, free association, and dream analysis.

4.2.2.1 Hypnosis

During the period from 1880–1882, Josef Breuer, of Vienna—and later his student, Freud—became famous for developing 'the talking cure' (that is, psycho-analysis) and using it successfully in the case of psycho-analysis's most famous and founding patient, 'Anna O.' Henri Ellenberger, one of the most influential commentators on psychodynamic theory, says that from the time that Breuer and Freud published this case, in 1895, "Anna O's story was given as the prototype of a cathartic cure and one of the basic occurrences that led Freud to the creation of psychoanalysis" (1972: 267).

Anna O. was the pseudonym given to a 21-year-old patient ("from a comfortable middle-class home," Kaplan 2004: 62) who had "fallen ill of a severe hysteria while she was nursing her sick father" (Freud 1923a, in *Complete Works*, p. 3912). For about the next two years, she displayed both psychological and physiological symptoms severe enough to warrant serious concern. She suffered from [i] "a rigid paralysis, accompanied by loss of sensation," in her right arm and leg; she experienced [ii] various visual disturbances; she was [iii] unable to swallow food or even water ("in spite of a tormenting thirst"). Even more alarming, "she was subject to [iv] conditions of 'absence'" (Freud 1910, in *Complete Works*, p. 2200), confusion, and delirium which were so severe that they appeared to alter her whole personality. Afterwards, she had no memory of these episodes.

No physical cause for these symptoms could be discovered. Freud noted that, when such symptoms were observed (especially in young women), and could not be accounted for by any evident physical pathology, the tendency among medical practitioners was to diagnose 'hysteria'. In Freud's time, it was assumed that "[m]edical skill is in most cases powerless against severe diseases of the brain; but neither can the doctor do anything against hysterical disorders" (Freud 1910, in *Complete Works*, p. 2201). Because the situation of the hysterical patient seemed to offer little if any scope for the application of doctors' medical skills, many nineteenth-century medical men were unsympathetic towards these kinds of cases.

But Breuer treated his patient with "both sympathy and interest" (Freud 1910, in *Complete Works*, p. 2202), even though he struggled to see what use his medical skills might be in the case. Although listening empathetically, with patience, and over an extended period of time, is standard behavior for contemporary psychotherapists, "[i]t should be borne in mind that Breuer's contemporary, the great Charcot, around this time began treating hysteria by having his patients ingest iron before hanging them from the ceiling in iron harnesses" (Middleton 2004: 242). Fortunately, in addition to her range of symptoms, Breuer's patient also possessed a remarkable degree of insight. "Following a hint given him by the patient herself [Breuer] put her into a state of hypnosis" and had her describe "the moods and thoughts that were uppermost in her mind" (ibid.). In this way, the two of them discovered the cause of Anna O.'s problems: she had been severely traumatized by this difficult period in her life, when she was caring for her sick father. On closer examination, the causes of her apparently unrelated symptoms became clear:

1. On one occasion, when sitting with her father, waiting for the doctor to arrive, Anna O.'s arm (which was hanging over the back of the chair) had gone to sleep. Halfway between sleep and wakefulness, Anna O. 'saw' a black snake coming to bite her father; she tried desperately to ward off the snake, only to find, to her horror, that her arm was paralyzed. Small wonder, perhaps, to find the same thing happening in her waking life (that is, when she found that she was unable to use her arm) that had so traumatized her when she dreamed about it.
2. On another occasion, Anna O. was again sitting at her father's bedside, when he asked her the time. In spite of bringing the watch very close to her eyes, she could not make out the time, on account of the unshed tears filling her eyes. The face of the watch, seen from so close, appeared very large—so again, it was unsurprising to find that she developed not only rather a squint, but also a condition called 'macropsia,' in which the patient experiences objects as being larger than they are in reality (and in which the patient herself feels as if she is much smaller than she really is, in proportion to the overlarge perceptual objects).
3. Seeing that it was summer, and very hot, it seemed unaccountable that the patient should develop a literal inability to drink water, in spite of suffering from severe thirst. She would pick up a glass, as if to drink—but as soon as the glass touched her lips, she would push it violently away again. But this symptom was immediately intelligible when it emerged, during her hypnosis, that Anna O. was unconsciously recalling an occasion when she had gone into a room where her friend was, and how that friend's 'horrid' little dog had drunk from her glass of water. Anna O. had tried to conceal her disgust, but the incident had

affected her profoundly—and that was found to be the reason for her own inability to drink.

4. "Finally, she was subject to conditions of '*absence*,' of confusion, of delirium, and of alteration of her whole personality" (Freud 1910, in *Complete Works*, p. 2200). When she was in her normal state, she had no recollection of these 'absences,' and no idea what might have caused them. However, during hypnosis, Breuer was able, through careful questioning, to deduce that during her 'absences' Anna O. was reliving the ordeal of her father's illness, her own illness which prevented her from continuing to care for him, and his subsequent death. She was not 'conscious' of these memories, and yet the memories overwhelmed her so completely that she temporarily lost track of what was really going on.

4.2.2.1.1 Hypnosis and the Unconscious

As Anna O.'s unconscious revealed (during hypnosis) more and more buried memories from this traumatic period, and as she talked them through with Breuer, her symptoms gradually faded, until eventually "she returned ... to a normal mental condition" (ibid.). Breuer/Freud concluded that "hysterical symptoms are residues (reminiscences) of profoundly moving experiences, which have been withdrawn from everyday consciousness" (Freud 1913, in *Complete Works*, p. 2545). Or rather, the mind has *attempted* to withdraw the memories from consciousness; but this attempted withdrawal has not been completely successful. The relevant memory is no longer accessible in conscious introspection, but it continues to exist, and to exert a pull on the rest of the psychic apparatus.

During hypnotherapy, the analyst helps to direct the patient's attention toward the ideas which have been imperfectly withdrawn from (introspective) consciousness, thus allowing the unconscious material to become fully 'conscious' again. Once it has become conscious, the unconscious material is largely deprived of "its pathogenic power" (ibid.). But the fact that the unconscious material had that pathogenic power in the first place shows that "an unconscious idea is not necessarily a weak one"

(Freud 1912, in *Complete Works*, p. 2579): it can determine not only what-it-is-like/experience, but even physical health or illness.

4.2.2.2 Repression and Free Association

Although it was hypnosis which had led to the 'cure' of Anna O., Freud viewed hypnosis as "a temperamental and, one might almost say, a mystical ally" (1910, in *Complete Works*, p. 2211), because only a minority of patients responded well to hypnosis. Therefore, Freud realized, it would be beneficial if he could achieve a similar result by working with the patients in their normal waking state. Although this initially seemed like "a senseless and hopeless undertaking," Freud remembered one of his teachers demonstrating that people could be put into "a state of hypnotic somnambulism," in which they "had all kinds of experiences" (ibid.). These patients denied being able to remember anything about what they experienced in these states, but Bernheim found that if he insisted that they could remember, and encouraged them to keep trying, they could eventually recall the experiences they claimed to have forgotten.

Accordingly, this is at first what Freud did with his patients. He got the patients to start talking about the circumstances in which the symptoms had arisen, and how they felt about the events that were taking place—effectively, to articulate their 'stream of consciousness'—up until he "reached a point with them at which they maintained that they knew nothing more," and the 'stream' was broken. At this point (with the case of Bernheim firmly in mind, no doubt), he insisted that the patient in fact *did* know more. He even went so far as to tell her that she would recall the forgotten memory at the exact moment when he placed his hand upon her forehead. This technique was moderately successful, but it was "a laborious procedure" (1910, in *Complete Works*, p. 2211), which was not ideal for long-term usage; moreover, there were many cases in which simple insistence was no help at all.

4.2.2.2.1 Repression

Although Freud ultimately abandoned the 'simple insistence' technique in favour of 'free association,' that was not before he had made an important observation. Very often, although patients declared that they could not remember something, their memories were really intact, and could be accessed when the patients tried hard enough to make the recall, having progressed closer and closer to the 'forgotten' memory by way of thinking about similar or related things. Freud concluded that there must be some 'force' preventing certain memories from becoming conscious; and to this force he gave the name Resistance.

But if there was some force (of Resistance) acting in the patients to prevent certain memories from getting out of the unconscious part of the mind, Freud reasoned, there must originally have been another force that removed these memories from the conscious part of the mind in the first place. "I gave the name of repression to this hypothetical process, and I considered that it was proved by the undeniable existence of resistance" (1910, in *Complete Works*, p. 2212). Resistance and Repression are the key to Freud's view of what was taking place in his hysterical patients.

4.2.2.2.2 Free Association

To do 'free association,' the patient is told to put herself "into a state of quiet, unreflecting self-observation, and to report to us [her] feelings, thoughts, memories in the order in which they occur to [her]" (Freud 1916–1917, in *Complete Works*, p. 3362). The rationale for this procedure was Freud's inability to believe "that an idea produced by a patient while [her] attention was on the stretch could be an arbitrary one and unrelated to the idea we were in search of." Rather, he assumed, "[t]he idea occurring to the patient must be in the nature of an *allusion* to the repressed element, like a representation of it in indirect speech" (1910, in *Complete Works*, p. 2217). For this reason, it was essential for patients doing free association to express immediately and unthinkingly whatever ideas happened to pop into their heads—"even if they felt objections to doing so, if, for instance, the thoughts seemed too disagreeable, too senseless, too unimportant or irrelevant" (1924, in *Complete Works*, p. 4105).

Using free association, Freud was able to achieve good results on a far larger number of his hysterical patients than he had by using hypnosis alone. The process of articulating whatever spontaneously came into her mind seemed to help bring unconscious content into consciousness; when the patient became conscious of the relevant unconscious psychological material, her symptoms disappeared. Similar results have been obtained more recently as well. For example, Pennebaker demonstrated that if emotions are aroused by writing about or discussing painful experiences, immune functioning, physical health, and adaptive behavior all improve (1989, 1997, in Westen 1999: 1082). Moreover, "[i]n the short run, expressing unpleasant emotion leads to a momentary increase in arousal [that is, 'stress'], but over the long run, doing so *decreases* arousal" (Pennebaker 1997, in Westen 1999: 1083). Acknowledging and expressing painful emotions was the whole point of Breuer/Freud's 'talking cure'—except their additional insight was that those painful memories were not just 'unconscious' in the innocuous 'preconscious' sense: they were *dynamically* unconscious. They were deeply and permanently unavailable to the patient in introspection, and could only be accessed through the use of the psychoanalytic techniques.

Although it is known by the name of 'free association,' this technique actually relies on the assumption that the associations made by the patients will *not* be 'free'—that is, they will be *determined* (for example, Lapsley and Stey 2011: 3) by the content which has been (unconsciously) repressed. Therefore, an unconscious idea is not necessarily a weak one: it can affect one's ability to recall other/related ideas.

4.2.2.3 Dream Analysis

Famously, Freud said that dream-interpretation was "the royal road to a knowledge of the unconscious" (1910, in *Complete Works*, p. 2220, also compare: 1912, in *Complete Works*, p. 2582). The original psycho-analysis practised by Freud was purely a method of treating pathological symptoms; but the use of dream analysis allowed the psychoanalytic framework to be applied even to the psychology of 'normal' people (for example, 1926, in *Complete Works*, p. 4404).

Freud said that the first thing to consider, regarding dreams and their significance, is *what is* a dream: that is, what are the essential characteristics of dreams, and in what ways do dreams differ from other psychological processes? What he determined was that "[t]he first thing common to all dreams would seem to be … that we are asleep during them." For this reason, clearly, it is necessary to understand exactly what sleep is, what its characteristics are, and what purposes it serves.

Freud asserts that

> [s]leep is a state in which I want to know nothing of the external world, in which I have taken my interest away from it. I put myself to sleep by withdrawing from the external world and keeping its stimuli away from me. I also go to sleep when I am fatigued by it. So when I go to sleep I say to the external world: 'Leave me in peace: I want to go to sleep.' … The biological purpose of sleep seems therefore to be rehabilitation, and its psychological characteristic suspense of interest in the world. (1916, in *Complete Works*, p. 3190)

In which case, it appears that dreams cannot be part of the rehabilitative function performed by sleep: on the contrary, they appear necessarily to be "an unwelcome addition" (Freud 1916, in *Complete Works*, p. 3191) to sleep—and this seems to be borne out by our assumption that to sleep without even dreaming is more restful than to have a sleep disturbed by any dreams at all.

4.2.2.3.1 Stimulus Dreams

But according to Freud, "*dreams are not disturbers of sleep*, as they are abusively called, but *guardians of sleep which get rid of disturbances of sleep*" (1916, in *Complete Works*, p. 3224). Sometimes, it is immediately evident what the stimulus is that threatens to disturb the sleeper's rest. Mourly Vold, Freud says, recorded three such dreams:

> He was given some eau-de-cologne to smell in his sleep. He dreamt he was in Cairo, in Johann Maria Farina's shop, and some further absurd adventures followed. Or, he was pinched lightly on the neck; he dreamt of a mustard plaster being applied to him and of a doctor who had treated him as a child. Or again, a drop of water was dropped on his forehead; he was in Italy, was sweating violently and was drinking white Orvieto wine. (1916, in *Complete Works*, p. 3193)

Similarly, one's ringing alarm clock can often be incorporated into one's dreams, in many different guises—from ringing church bells, to jingling sleigh bells, even to shattering crockery (for example, 1916, in *Complete Works*, p. 3194). In these cases, it is very clear that the ringing alarm-clock caused each of the dreams—but what is not clear is why the ringing sound was incorporated into the dream in a particular form (different in every case); or even, whether the form taken by the stimulus in the dream had any meaning at all. Freud thought that the actual form taken by these stimulus dreams was probably not relevant to any important unconscious psychological processes. In order to investigate how unconscious psychological processes contribute to dreams, Freud realized, he would have to look for dreams which were not elicited by discernible physical stimuli.

4.2.2.3.2 Children's Dreams

Freud found that the ideal kind of dream to analyse, in order to find out more about unconscious psychological processes, was a certain kind of dream found children: dreams which are short, clear, coherent, easy to understand and unambiguous.

Freud gives three examples of these dreams.

(a) A boy of 22 months was told to hand over a basket of cherries to someone as a birthday present. He was obviously very unwilling to do it, although he was promised that he should have a few of them for himself. Next morning he reported having dreamt: '*Hermann eaten all the chewwies!*'

(b) A girl of 3¼ years was taken across the lake for the first time. At the landing-stage she did not want to leave the boat and wept bitterly. The crossing had been too short for her. Next morning she announced: '*Last night I went on the lake.*' We may safely add that this crossing had lasted longer. (1916, in *Complete Works*, p. 3222)

And finally

(c) A boy of 5¼ years was taken on an excursion up the Echerntal near Hallstatt. He had been told that Hallstatt was at the foot of the Dachstein. He had shown great interest in this mountain. There was a fine view of it from where he was staying at Aussee, and the Simony Hut on it could be made out through a telescope. … The excursion began in an atmosphere of cheerful expectation. Whenever a fresh mountain came into view the boy asked 'Is that the Dachstein?' and he became more and more depressed the more often he was told it was not. Finally, he fell completely silent and refused to go with the rest of the party up the short ascent to the waterfall, and it was thought that he must be overtired. But next morning he said with a radient face: 'Last night I dreamt *we were at the Simony Hut.*' So that had been what he expected to do on the excursion. (Freud 1916, in *Complete Works*, p. 3222)

Freud discovered that an overwhelming number of other children's dreams seemed to take this same form. The main conclusions he drew from analysing the children's dreams were as follows:

1. It is not necessary to be an expert on psycho-analysis in order to understand these dreams fully: something from the child's experience in the day preceding the dream explains the dream. Such dreams, far from being confused and meaningless, are "intelligible, completely normal mental acts" (1916, in *Complete Works*, p. 3222). This supports Freud's claim that (introspective) consciousness is a quality of only *some* psychical phenomena, because the process which produces dreams can comfortably be called 'psychical,' in spite of not being 'conscious.'

2. Although children's experiences from the day preceding the dream are involved in the production of the dreams, it is only a very specific type of experience which causes dreaming: "an experience of the previous day, which has left behind it a regret, a longing, a wish that has not been dealt with" (1916, in *Complete Works*, p. 3223). We can say that, just as simple stimulus-dreams protect the dreamer's sleep from being disturbed by physical stimuli, so these children's dreams protect the children's sleep from being disturbed by 'mental stimuli.'
3. Typical children's dreams appear to be straightforward—but there is one respect in which they always differ from the memories from which they are derived: "*The dream produces a direct, undisguised fulfilment of that wish*" (ibid.). In the real world, the child's wish was not fulfilled, which is why its unconscious mind went to work, during sleep, to attempt to achieve fulfilment of the wish. Just like free associations, dreams are the furthest possible thing from being arbitrary: they are motivated phenomena, determined by important unconscious psychological material.

4.2.2.3.3 Dream Distortion

That all seems reasonable enough, and, as Freud notes, "[i]f dreams of the infantile kind were the only ones, the problem would be solved and our task finished" (1916, in *Complete Works*, p. 3227). Unfortunately, however, the situation is not so simple: "the content of adults' dreams is as a rule unintelligible and could not look more unlike the fulfilment of a wish" (1916, in *Complete Works*, p. 2222). In adults' dreams, the underlying wish has undergone "a far-reaching distortion"; and Freud suspects that "to explain this distortion we shall need the psycho-analytic technique" (1916, in *Complete Works*, p. 3227).

The first step in understanding dream-distortion is to distinguish the surface appearance (of what the dream was about) from "the *latent dream-thoughts*, which you must suppose were present in the unconscious" (1916, in *Complete Works*, p. 2222). The latent dream-thoughts are of the same kind as ordinary conscious thoughts: although they are not properly 'conscious,' they are unconscious only in the (temporary) *preconscious* sense. But during the state of sleep, when the normal ('conscious') level of control is relaxed, they come into contact with some repressed material. The result is that both the conscious thoughts about the day *and* some repressed content combine to produce the manifest dream content.

In 'the manifest dream,' some of the elements of 'the latent dream' may have been reproduced faithfully, but others may be missing altogether, modified slightly, or modified so drastically as to be unrecognizable. The process which distorts the psychological material into the unrecognizable form presented in the manifest content of the dream is what Freud refers to as the 'dream-work.' "It deserves our closest theoretical interest, since we are able to study in it, as nowhere else, what unsuspected psychical processes can occur … between two separate psychical systems like the conscious and unconscious" (Freud 1910, in *Complete Works*, p. 2223).

As an example of dream distortion, Freud recounts the dream of "a highly-esteemed and cultivated elderly lady" (1916, in *Complete Works*, p. 3232). The dream concerns "'love services' in war time" (the German *Liebesdienste* "means in the first instance 'services performed for love,' i.e. 'unremunerated services'; but it could bear another, less respectable, meaning"):

> *She went to Garrison Hospital No. 1 and informed the sentry at the gate that she must speak to the Chief Medical Officer … as she wanted to volunteer for service at the hospital. She pronounced the word 'service' in such a way that the N.C.O. at once understood that she meant 'love service'. Since she was an elderly lady, after some hesitation he allowed her to pass. Instead of finding the Chief Medical Officer, however, she reached a large and gloomy apartment in which a number of officers and army doctors were standing and sitting round a long table. She approached a staff surgeon with her request, and he understood her meaning after she had said only a few words. The actual wording of her speech in the dream was: 'I and many other women and girls in Vienna are ready to…' at this point in the dream her words turned into a mumble '…for the troops—officers and other ranks without distinction.' She could tell from the expressions on the officers' faces, partly embarrassed and partly sly, that everyone had understood her meaning correctly. The lady went on: … there's one condition that must be observed: age must be respected. It must never happen that an elderly woman … (mumble) … a mere boy. That would be terrible.'* (1916, in *Complete Works*, p. 3233)

For the psycho-analyst, what is even more interesting about this dream than what actually happens in it is the fact that it has some significant gaps. Freud likens the suppression of these 'objectionable' parts of the dream to what takes place when a press censor analyses a political newspaper. "[Y]ou will find that here and there the text is absent and in its place nothing except the white paper is to be seen. … In these empty places there was something that displeased the higher censorship authorities and for that reason it was removed—a pity, you feel, since no doubt it was the most interesting thing in the paper" (ibid.). At other times, however, the article's writer has anticipated which passages the censor might object to, and has toned them down in advance. Then there are no blank spaces where the censor has actually removed text, but the censor's influence has still affected the final product, insofar as the writer has had to weigh her words more carefully than she might have otherwise. In just the same way, a kind of 'dream-censorship' is responsible for gaps and distortions in the manifest content of dreams.

This censorship embodies "the dreamer's waking judgment" (Freud 1916, in *Complete Works*, p. 3237); that is, the motives and considerations which the censorship process seeks to distort are precisely those that do not correspond to the values the dreamer consciously espouses. This process of distortions appears to be identical to the unconscious cognitive process which I said took place in 'the counter-attitudinal advocacy experiment' [Sect. 3.2.3.1]. In that experiment, it was found that if the participants had good reasons for writing an essay expressing views which were contrary to their actual views (for example, if they were bribed), their actual attitudes did not change—but if they were given a *bad* reason for writing the essay, they *unconsciously* modified their own views to make them more similar to the views they were expressing in their essays, so that the conflict would be reduced. Similarly, according to Freud's view of dream distortion, the dreamer's own mind *unconsciously* disguises the real meaning behind the dream, in order to protect the dreamer against having to acknowledge any uncomfortable truths.

Thus, we could probably say that the main insight to be obtained from the consideration of the dream work is the same one to be obtained from the consideration of hypnosis and free association: "The more we seek to win our way to a metapsychological view of mental life, the more we must learn to emancipate ourselves from the importance of the symptom of 'being conscious'" (Freud 1915, in *Complete Works*, p. 3014). The unconscious material which is revealed by dream analysis is dynamically unconscious, and completely unavailable to introspection, but nevertheless it profoundly influences what-it-is-like/experience. Therefore, an unconscious idea is not necessarily a weak one: it can determine the form taken by one's dreams.

4.2.2.4 The Lesson (of 'Psychoanalytic Techniques') for Philosophy

The psychological disorders Freud encountered in his patients provide evidence that some unconscious ideas cannot be made 'conscious' (by means, for example, of having higher order thoughts directed at them [compare: Sect. 3.2.1.1], or of having attention paid to them [compare: Sect. 3.3.2]) by the patient herself. But some of this psychological material, although it is totally inaccessible to introspection, can have powerful effects not only on the patient's subjective (psychological) what-it-is-like/experience, but even on her physical body. The use of the psychoanalytic techniques is effective in bringing those dynamically unconscious contents back into the ambit of introspective consciousness, and this is what deprives those contents of their pathogenic power [Sect. 4.2.2.1]. The postulation of unconscious psychological material is not only *legitimate*, but also *necessary* [Sect. 4.2.1.3], in order to make sense of these patients' cases.

Although, as I explained [Sect. 4.2.2.2], it is repressed (/dynamically unconscious) material which leads to psychopathologies, and therefore illuminating these contents is what leads to the patient's cure, "none of this implies that the quality of being conscious has lost its importance for us. It remains the one light which illuminates our path and leads us through the darkness of mental life" (Freud 1940b, in *Complete Works*, p. 5071).

4.2.2.4.1 *Recovering Dynamically Unconscious Material*

Importantly, although dynamically unconscious content may never *in practice* come to be known introspectively in consciousness, it is always "*in principle* dynamically recoverable" (Marshall 2000: 46, my emphasis)—perhaps with the help of a trained psycho-analyst. If dynamically unconscious material were not accessible to (introspective) consciousness even in principle, then perhaps some of these charges about Freudian theory being impervious to empirical investigation would be justified:

> 'Unconscious phantasies can … never be perceived directly; they can … only be *inferred* from the data of conscious perception' (Solms 2003, p. 103, his italics). Talvitie (2009) correctly notes here that this would make a science of the unconscious impossible, and it is further unclear how either Solms, Freud or anyone else could make knowledge claims of anything that is literally unknowable. (Boag 2015: 256)

But fortunately it is not the case that unconscious material is inaccessible even in principle. The existence of dynamically unconscious material is corroborated not only every time it makes theoretical sense of a patient's symptoms to postulate hypothetical unconscious content, but also (even more) every time helping the patient to become aware of that unconscious content is effective in alleviating the patient's symptoms.

4.2.3 (Psychological) What-It-Is-Like/Experience and (Introspective) Consciousness: Conclusion

Although, in Freud's time, it seemed absurd to almost everybody to suggest that something 'mental' was not necessarily also 'conscious,' we could now probably agree that "the shoe is on the other foot and the contemporary challenge is to demonstrate that there are causally efficacious, proximate *conscious* intentions" (Pataki 2015: 22).

As I explained, Freud was well aware both that 'the brain is the bodily organ of mental life,' and that 'our acts of consciousness' present the appearance of being different in kind from physical bodies. The two apparently irreconcilable facts which Freud was concerned with reconciling are essentially the same irreconcilable facts that leave us with the illusion of the explanatory gap; and Freud was quite correct to insist (just as I have been insisting) that the true essence of what is psychical—whether 'conscious' or 'unconscious'—should be sought in physical processes.

Freud's clinical cases indicate that psychological material which is unknown to the individual (and cannot be accessed through introspection) can significantly impact subjective, psychological what-it-is-like/experience. Freud's theoretical framework of the structure of consciousness helps make conceptual sense of this.

Because the existence and significance of dynamically unconscious material can be both theoretically accounted for [Sect. 4.2.1] and empirically corroborated [Sect. 4.2.2], we should be careful not to conflate (psychological) what-it-is-like/experience with (introspective) 'consciousness.'

Freud vs Brentano:

According to Wakefield, it is possible to discern, "scattered throughout Freud's writings … a sophisticated and subtle philosophical argument about the nature of the mental" (2018: 2). But as I argued [Sect. 4.2.1.1], Freud, as a genuine physicalist (as opposed to a closet dualist, Sect. 2.2.3), could not even acknowledge *the existence of* such a thing as 'the mental.' Speaking about 'the mental' as having a certain 'nature' presupposes that 'the mental' *exists*; and recognizing that 'the mental' exists (and/or is different in kind from 'the physical') is constitutive of dualism. Therefore, arguing about 'the nature of the mental' would be in direct conflict with Freud's conviction that everything that exists is physical, which is why I am highly sceptical about Wakefield's assertion that Freud has done so.

Another focus of Wakefield's assessment of Freud that leads me to suppose that he may not have understood Freud correctly is his suggestion that Freud's sophisticated and subtle argument has anything to do with the philosopher Franz Brentano. Wakefield writes that "we know from Freud's letters that his philosophy teacher, Franz Brentano, instilled in Freud some of his extreme negative attitudes toward earlier philosophers even as he inspired Freud's love of philosophy" (2018: 4). Wakefield even goes so far as to claim that "Freud's disdain for philosophy was a would-be philosopher's disdain for an admired subject being poorly pursued."

Finally, he refers to "Freud's supposed negative attitude toward philosophers and philosophy," which, he claims, "has been presented so many times that I do not go through it again."

But in the first place, I am concerned that looking for Freud's most mature and well-considered views in his personal letters, rather than in his published works, may not be advisable. In Freud's correspondence with his own friends, there were likely many shared understandings of favourite topics of conversation which an outsider, reading the letters many years later, would be virtually guaranteed to overlook. In the second place: whether or not Freud felt 'disdain for an admired subject being poorly pursued' is similarly not something Wakefield (or any other contemporary Freud reader) is in a position to speculate about either on the basis of Freud's published works or on the basis of his personal correspondence. Therefore, whether or not Freud was known to have admired philosophy should not be allowed to influence any careful reading of what Freud *actually said* about philosophy. And in fact, as I have already mentioned, what Freud actually said was that, in order to shed any light on 'the psychical,' '*the philosophers at least* were obliged to assume that there were organic processes parallel to the conscious psychical ones' [Sect. 4.2]. In contrast both to many professionals and many laypersons among Freud's contemporaries, 'the philosophers' shared Freud's own view, that psychical processes must be similar in kind to all other natural processes—in which case, it would probably be sensible *not* to suppose that Freud had a negative attitude towards philosophers.

Wakefield thinks that it was owing to "Freud's deep respect for Brentano that … Freud never mentioned Brentano by name as a target of criticism throughout his works" (Wakefield 2018: 169)—but I wonder, would it not perhaps be closer to the truth to say that Freud never mentioned Brentano by name simply because *Freud did not consider Brentano's views to be relevant to the aims of psychoanalysis?* Boag (2015: 244) notes that "Brentano (like Descartes) denies the possibility of unconscious mental acts"—but this may be a good reason for thinking Brentano's views may not have had much influence on Freud. We already know that Freud was flying in the face of the prevailing 'Cartesian' view that equated 'mental' with 'conscious' [Sect. 4.2]—so would it not be more likely that, if anything, Freud was reacting against Descartes' views, rather than Brentano's views?

According to Boag, Brentano took "mental acts" to be "self-reflexively known" (ibid.)—and as his example he says "For example … in the case of hearing a sound, both sound and the act of hearing the sound are known simultaneously." This he takes to be incompatible with Freud's view that many important psychological processes are unconscious. But in the first place, as I have argued at length in Sect. 3.2, a large part of perception (of sounds, as well as other things) takes place *without* 'consciousness'—which means that neither 'the sound' nor 'the act of hearing' would be 'known.' In the second place, as I have also argued at length [Sect. 3.3], the subjective (psychological) what-it-is-like/experience Freud was interested in is dissociated from perceptual (or cognitive) 'consciousness.' Here, then, is another reason for being skeptical of the claim that Freud was significantly influenced by Brentano's views—or indeed that knowing about Brentano's views is at all relevant to understanding Freud's.

According to Pataki (2015: 9), in fact, "there is no evidence in his writings that Freud possessed anything like a *scholarly* knowledge of … Brentano"—although, seeing that the "popularized ideas" of Brentano and other Enlightenment scholars were very much "in the air" during this period, "so large a mind as Freud's could scarcely have failed to assimilate them." Moreover, any portrayal of Freud as "concerned explicitly to counter Brentano's denial of unconscious mentation" can be based on nothing more than "speculations based on superficial thematic affinity" (Pataki 2015: 42 [Note 8])—seeing that speculation is all anyone can do in this case, in the absence of more concrete evidence.

Boag writes "It is to Freud's detriment [sic] that he was not more philosophically minded, and one can agree … that it is unfortunate that Freud did not fully engage with Brentano's thesis of Intentionality when considering his formulation of mentality, conscious or otherwise" (2015: 239). But actually—seeing that, as I have repeatedly explained, the only people who can say anything about 'the mental' are dualists, whether self-professed- or closet—I think it would be much more plausible to say that the fact that Freud did not engage with Brentanian Intentionality only serves to confirm his commitment to physicalism. Whereas these (closet dualist) philosophers talk about such concepts as 'intentionality' and 'the mental,' it seems significant to me that Freud tends to use the term 'the mental *apparatus*,' rather than just 'the mental' (he sometimes says 'the psychical,' but 'the psychical' and 'the mental' are plausibly not the same thing). This usage is in keeping with his claim that psychical processes are probably similar in kind to all other natural processes [Sect. 4.2]. For Freud, 'the mental apparatus' is a system of neural mechanisms, along with the system of subjective, psychological experiences which are both produced by, and in a causal sense indistinguishable from, this system.

4.3 (Psychological) What-it-is-like/experience and Rationality

I explained [Sect. 4.2] how common it is to take (psychological) what-it-is-like/experience to be the same thing as 'consciousness'; and I said that another common mistake was to take (psychological) what-it-is-like/experience to be the same thing as—or at least, to be a question of—'rationality' (Chap. 4, para. 3). Writers like Churchland and Chalmers were able to think of 'the mind' much like a computing machine which performs logical operations [Sect. 2.3], and this was largely because they assumed that 'mental processes' were the kinds of thing which consisted of informational input, the application of logical (computational) rules, and (therefore) the logically appropriate output.

In this section, I am first [Sect. 4.3.1] going to take a brief look at the main problem inherent in trying to give any sort of 'account' of 'irrationality' at all. Then I will examine Freud's model of 'the structures of the mind'; in particular, I will discuss how using this model helps to account for logically incompatible ('irrational') motives. Secondly [Sect. 4.3.2], I will discuss, in some depth, a particular article by Mark Solms which the author thinks reveals some serious flaws in this 'structures of the mind' model.[6] By defeating Solms' challenge to Freud's framework, I hope substantially to strengthen the claims of that framework—a framework which, as I have explained, rests neither on the assumption that (psychological) what-it-is-like/experience is a question of 'consciousness,' nor the assumption that it is a question of 'rationality.'

4.3.1 Accounting for Irrationality

As I said [Sect. 2.3.1], 'philosophical behaviorism' is the view that our habitual practice of characterizing what people do in terms of 'mental predicates' is not really the practice we think it is. Ryle claimed that when we use such terms, rather than making 'untestable inferences to any ghostly processes occurring in streams of consciousness which we are debarred from visiting,' what we are actually doing is describing the 'predominantly public behavior' of the individual [Sect. 2.3.1.2]. On this view, "statements containing mental terms can be translated, without loss of meaning, into subjunctive conditionals about what the individual will do in various circumstances" (Tanney 1949: xxv).

According to behaviorism, therefore, it must be assumed that rationality is a fundamental characteristic of human persons. According to both philosophical behaviorists like Ryle, and behaviorist psychologists [Sect. 4.1], 'mental predicates' can be understood purely in terms of an individual's dispositions to behave in certain ways in certain situations—and having a disposition to behave in a certain way in a certain situation *only makes sense in the context of rational individuals.*

For example, rather than saying that an individual *feels afraid*, the philosophical behaviorist would say that, all things being equal, under these circumstances, this individual would tend to run away. But for this to make sense, the individual in question must have a capacity for rationality. The individual needs to be able to understand that (i) that thing causing the fear constitutes a threat to the individual's safety, and that (ii) running away constitutes a good means of neutralizing that threat. Accordingly, although most (normal) humans are 'rational' in the requisite sense, very young children, for example—or individuals with various types of cognitive deficits—would not be members of the community of rational individuals whose behavior would be of interest to any variety of behaviorist.

Many contemporary philosophers (for example, Marcia Cavell, Donald Davidson) also hold views according to which rationality is an essential prerequisite for our capacity for conscious experience (Snelling 2000: 102). On Davidson's view, mental states can be ascribed to persons only if the ascription of those states to those persons enables an external observer to make sense of their behavior. And a third-person observer would only be able to make sense of the behavior insofar as that behavior was, in at least some minimal sense, 'rational.' This position is similar to the behaviorist view, insofar as "what is the case about inner states or contents begins and ends with what can be attributed in interpretation; there is no further matter to be considered, no irreducibly *mental* content" (ibid.). Cavell refers to the view as 'externalism.' According to externalism, a capacity for rationality is a prerequisite for being able to find the world intelligible or meaningful at all.

But if persons are essentially rational, it is difficult to make sense of what it might mean for them to behave in an *ir*rational way—which, it seems safe to say, they sometimes do. The problem is that any suggestion that 'irrational' behavior can be *explained* is tantamount to saying that it is actually *not* irrational at all. As Davidson says, "if we explain it too well, we turn it into a concealed form of rationality" (1982, in Marshall 2000: 46). The whole point of 'irrational' behavior is that it *cannot* be 'explained' (in terms of reasons). This is a problem for many accounts of irrationality: they effectively explain irrationality out of existence.

In contrast to behaviorist and/or 'externalist' views, internalist views "see mental content as essentially *private*" (Snelling 2000: 102). The psychodynamic view—as well as, for instance, Descartes' view—is an example of internalism; such views are therefore unacceptable to externalists like Cavell and Davidson. But I am very tempted to conclude that the real problem here is not to try and determine whether externalism or internalism is a better account of irrationality. I argued [Sect. 2.2] that not only were dualism and materialism deeply problematic, but both were both completely unsuited to accounting for the explanatory gap; similarly, here, I think I would want to argue that, because internalism and externalism are both so problematic, neither is any use for producing a philosophical account of 'irrationality.'

Philosophers who take a "more liberal" position on the question of rationality suggest that any account which is based on the assumption that humans are essentially rational will always have *gaps* in it: "gaps which occur because of the existence of various sorts of irrational behavior which are resistant to common-sense psychological explanation, yet bear features which cry out for interpretation through an *extended* psychology of motive" (Snelling 2000: 102–103). That seems plausible: accounts which rely on the assumption of rationality will always be incomplete, because they are unable to make sense of 'irrationality'—which nevertheless does seem to be a genuine part of human experience.

4.3.1.1 The Psychodynamic Account of Irrationality

Tamas Pataki thinks that even if Freud was only approximately right about the conscious and unconscious psychological forces he proposed, then some of the core tenets of psychodynamic theory should take a central place within any philosophical framework with any pretensions to explaining irrational action or belief formation. This is because "although one cannot desire what one knows to be impossible, one can certainly wish for it" (2015: 2, Note 5)—and such non-rational wishes can and do play a central role in the psychodynamic picture of motivation.

For Freud, logical conflict between incompatible drives is the source of almost all psychological problems. In psychodynamic theory, logically incompatible motives cannot be *conscious* simultaneously; however, since the Freudian view denies that 'the psychical' can be equated with 'what is conscious,' logically incompatible motives demonstrably can and do *exist* simultaneously (as long as at least one of those motives is unconscious). That is the whole point of the dynamic unconscious: the process of *repression* [Sect. 4.2.2.2] deals with logically incompatible (simultaneous) motives by forcibly removing at least one of those motives from consciousness, and relegating it to the dynamic unconscious. Therefore, on this view, there is no difficulty in saying both that persons are inherently rational, and that persons are often motivated by logically incompatible forces. In virtue of its denial that consciousness is the essence of the psychical, psychodynamic theory is also able to explain how 'irrationality' is able to proliferate in such apparently rational beings as human persons.

In Chap. 3, I explained how people tend (unconsciously) to re-evaluate/change their own attitudes, if they hold views which are in conflict with their overt behavior [Sect. 3.2.3]. In that chapter, I was using the cognitive dissonance studies to demonstrate that many cognitive operations are performed 'unconsciously'; but what they demonstrate equally well is the Freudian contention that, while it is not possible for somebody to be *conscious* of logically incompatible motives simultaneously, it is common for people to be simultaneously pulled in opposite directions by such conflicting motives, if at least some of the motives are unconscious.

In those cognitive dissonance studies, if people had no good extrinsic reason for shifting their opinions, they *unconsciously* shifted their own opinions towards the views they were expressing, in order that their actual own views would not conflict so strongly with the views they were expressing. In a very similar way, Freud's patients' minds unconsciously protected them against having to recognize that they were being motivated simultaneously by logically incompatible urges, by forcing at least one of these urges to be unconscious.

4.3.1.1.1 Clinical Considerations

As I explained [Sect. 4.2.2.1], (i) Anna O. experienced an inability to move her arm and leg, although there was nothing wrong with them physically (because of her repressed memory, with its associated sense of powerlessness, of that time when her forearm went to sleep while she was sitting at her sick father's bedside). (ii) Her vision was disturbed, even though there was nothing wrong with her eyes (because of her repressed memory of that occasion when her father asked her the time, and she was unable to see the clock properly, because of all the tears in her eyes). (iii) She found herself physically unable to swallow water, in spite of her severe thirst (because of her repressed memory of her companion allowing a little dog to drink from her own glass). And at times (iv) she appeared to be recalling experiences from her own past so intensely that she temporarily lost her awareness of the present. She had within herself simultaneously the desire to move and not to move; to see and not to see; to drink and not to drink; recall the past and not to recall the past.

As it turned out, Anna O. was by no means unique in this respect. Freud found that the process of repression always seemed to take the following form: after "a short conflict … the idea which had appeared before consciousness as the vehicle of this irreconcilable wish fell a victim to repression [and] was pushed out of consciousness" (1910, in *Complete Works*, p. 2212). When the idea was brought back to (introspective) consciousness during psychoanalysis, the patient became aware of the incompatibility of the two motives; the repressed material was thus deprived of "its pathogenic power," and the hysterical symptoms were eradicated. In this case, the relevant psychological material was 'unconscious' in the dynamic sense of having been repressed, and the patient's psyche unconsciously repressed this idea *for a very good reason*: it was logically incompatible with her conscious values.

4.3.1.1.2 Recent Studies

A range of studies from the 1990s demonstrated that repressed psychological material could have detrimental effects on various aspects of patients' health. For example, Dozier and Kobak found that people who used avoidant strategies while they recalled stressful or hurtful memories involving their parents showed not less but more physiological reactivity (that is, an indication of 'stress') during the task than those who did not use such strategies (in Westen 1999: 1081). Shedler, Mayman and Manis found that, when required to undergo a mildly stressful procedure (like reading aloud, story-telling, et cetera), subjects who showed signs of having been affected by stress relating to these memories, but who were themselves unaware of the effect the stress had had on them, actually "manifested considerably more cardiac reactivity on a measure related to heart disease" (ibid.) than either subjects with both unstressful early memories and unstressful (current) self-reports *or subjects with both stressful early memories and stressful current self-reports!*

This shows that the real stress was not to be found in what the patients themselves were aware of, or even in the repressed content, but in the *conflict* between their conscious and unconscious ideas. In these cases, the relevant (unconscious) ideas were not just innocuously unconscious in the sense of *pre*conscious; they were actively (dynamically) being repressed by psychological forces outside the patient's awareness. Moreover, they were being repressed *precisely because they were logically incompatible with the patient's overt beliefs*. In other words: some studies at the end of the twentieth century seemed to confirm what Freud had been saying since the end of the nineteenth.

All of which just goes to show that, insofar as it is more satisfactory to account for 'irrationality' in terms of conflicting conscious and unconscious motives than it is to try and account for it by giving reasons, philosophical (behaviorist and/or internalist/externalist) accounts of irrationality are problematic, and the psychodynamic account is preferable.

4.3.1.2 The Structures of the Mind

Within the framework of Freud's original 'psycho-analysis,' there are different 'points of view,' which Freud employed respectively at different stages of his career, from which to consider 'mental life.' The *dynamic* standpoint is concerned with "the interplay of forces" (Freud 1926, in *Complete Works*, p. 4403) within the mental apparatus; the *economic* standpoint is concerned with the quantitative amounts of energy which are dispersed throughout the mental apparatus during different types of psychological processes; and from the *topographical* point of view, "psycho-analysis regards the mental apparatus as a compound instrument, and endeavours to determine at what points in it the various mental processes take place" (1926, in *Complete Works*, p. 4404).

Initially, the topographical model was identical to the iceberg model [Sect. 4.2.1.3]; however, although Freud's discovery of the phenomenon of repression confirmed for him the necessity of dividing up consciousness into Conscious, Unconscious and Preconscious, he found that this "leaves us not at the end but at the beginning of a psychological theory"

(1910, in *Complete Works*, p. 2214). In the course of his practice of psycho-analysis, the model of the three levels of consciousness "proved to be inadequate and, for practical purposes, insufficient" (1923b, in *Complete Works*, p. 3951) for getting to grips with phenomena like repression; and he realized that, in order to articulate his new insights about the psychical apparatus, he needed some new conceptual tools.

With the publication of *The Ego and the Id* (in 1923), Freud completed his new formulation of the topographical view. In this final model (also known as the 'structural' model), each of the three structures of the mind is conceived of as an autonomous entity, and each one can be ascribed its own reasons and motives. The 'structural model' is particularly useful for describing how the mental apparatus deals with logically incompatible motives by ensuring that they are never conscious simultaneously.

4.3.1.2.1 The Id

In the structural model, the "ill-defined notions of unconsciousness"[7] (Lapsley and Stey 2011: 5) of the previous models are largely replaced by the notion of the *id*. The term *id*, which means 'it,' in German, refers to the 'oldest' "psychical province" (Freud 1940a, in *Complete Works*, p. 4957)—in the sense that it contains that part of the psyche which is passed down through many generations, rather than acquired by each individual. Freud says that very little is known about the sensations and feelings which are important to the id, but that the best examples of them are those to do with pleasure and unpleasure.

The id is 'unconscious,' in the sense that nothing can be known introspectively about the id, even by the person whose id it is. Only "through the medium of another agency" (1940a, in *Complete Works*, p. 5004)—her psychoanalyst, perhaps—can an individual gain knowledge of her own id. However (and significantly, as I will explain below, Sect. 4.3.2.2), although he described the id as 'unconscious,' Freud also said that the 'internal' perceptions of the id—primarily, feelings of pleasure or unpleasure—can still come about "even when consciousness is clouded" (1923b, in *Complete Works*, p. 3956).

Insofar as humans do still have biological drives or instincts, these can be found in the id. Because the 'world' of the id is the world of its own interior, and because it does not perceive anything at all in the external world, it detects these biological drives or instincts "with extraordinary acuteness" (1940a, in *Complete Works*, p. 5005). All instincts are aimed at instant gratification; by their very nature, they are the antithesis of rational consideration. Unsurprisingly, therefore, the id "knows no solicitude about ensuring survival"; and "[t]he processes which are possible … in the id … [are not] subject to the critical restrictions of logic" (1940a, in *Complete Works*, p. 5004).

Just as the dynamic unconscious was like the huge underwater part of the iceberg, on the topographical model, so the id forms the bulk of the psychic apparatus on the structural model. And the id categorically does not operate in accordance with rationality.

4.3.1.2.2 The Ego

Because the id does not concern itself with ensuring survival, if the id were the only structure of the mind, the survival of the race would be in jeopardy. Fortunately, in almost all cases, the *ego* is there to worry about "the task of self-preservation, which the id appears to neglect" (1940a, in *Complete Works*, p. 5005). The ego is the "coherent organization of mental processes … which supervises all its own constituent processes, and which goes to sleep at night, though even then it exercises the censorship upon dreams" (1923b, in *Complete Works*, p. 3951). Whereas the id recognizes only 'the pleasure principle' as a valid motivation, the ego operates in accordance with the *reality* principle. Its main job is to decide whether or not conditions are favourable for allowing the id to pursue its pleasures, whether to postpone this pursuit, or prohibit it altogether.

Because it is concerned with 'reality,' the ego relies heavily on different kinds of sense perceptions. We can say that, in the ego, "perception plays the part which in the id falls to instinct." Vision, hearing, smell, and taste are all useful for finding out about external reality—but in terms of the sense of *touch*, the body "yields two kinds of sensations" (1923b, in *Complete Works*, p. 3959). The first kind is the sensation of the object against the skin, in the sense in which this perception allows one to identify what that external object is like; and the second kind is the 'internal' perception of how it feels to be a sensate body touching something. This type of sensation is arguably even more strongly rooted in the body than sensory perceptions, since it has little or nothing to do with the external environment. The ego, in discharging its function of reconciling the urges of the id with the dictates of reality, has to deal extensively with both the body's 'external' perceptual sensations and its 'internal' sensations/feelings. That is why Freud said that the ego was "first and foremost a bodily ego" (1923b, in *Complete Works*, p. 3960), with respect to both these senses (of touch).

But although, for these reasons, the ego has strong ties to the physical body, it is also important to note that the ego (and in fact both the other 'structures of the mind' as well) is a *functional* entity. Freud used the term to refer to whatever physical and/or psychological structures and/or processes might contribute to what he viewed as the ego's main functional role: finding ways of reconciling the urges of the id with the constraints of reality. Another important thing to notice about the structural model is that, at the same time that the id and the ego are functionally distinct enough to have different motives, "one must not take the difference between ego and id in too hard-and-fast a sense, nor forget that the ego is a specially differentiated part of the id" (1923b, in *Complete Works*, p. 3972).

4.3.1.2.3 The Super-Ego

I said that the ego's main job is mediating between the id and the external world—but strictly speaking, this job is only necessary when an individual is in "the first period of childhood." Around the age of five, an additional layer of discipline is added to the structures of the mind: "[w]e call this agency the *super-ego* and are aware of it in its judicial functions as our *conscience*" (1940a, in *Complete Works*, p. 5010).

The super-ego (or 'ego-ideal') contains a pattern of the 'ideal' way for the child to be; and Freud said that behind the origin of this ego-ideal "lies hidden the first and most important identification of all, the identification with the father" (although he notes that, since young children do not appear to differentiate between the sexes, it might be better to say 'with the parents,' 1923b, in *Complete Works*, p. 3444). The superego can be thought of as an imprint left by the earliest impulses of a child's id, and particularly how the child was taught, by its father and mother, to exercise control over those impulses.

The structural view could be taken as a satisfactory functional account of the psychic apparatus (although Freud also had certain ideas about where in the brain these functions probably took place, as I will explain in Sect. 5.2.2.1). The id is responsible for the strong instinctive drives which motivate most of people's behavior; the ego is responsible for trying to reconcile these drives with the constraints of reality; and the superego is responsible for trying to balance the strong instinctual drives of the id not only with the constraints of reality but also with the constraints of an individual's moral beliefs.

Insofar as it is more satisfactory to account for 'irrationality' in terms of the conflicting motives of autonomous structures of the mind than by giving reasons, philosophical (behaviorist and/or internalist/externalist) accounts of irrationality are problematic, and the psychodynamic account is preferable.

4.3.1.3 The Lesson (of 'Accounting for Irrationality') for Philosophy

Strictly speaking, of course, it is not really necessary to appeal to something like the structural model to account for logically incompatible motives: the problem has already been solved, as soon as one proposes that individuals can be simultaneously motivated by (different) conscious and unconscious forces. As I explained [Sect. 4.3.1.1], there is no difficulty about suggesting that 'rational' beings like ourselves can be driven by incompatible motives simultaneously, so long as at least one of those motives remains unconscious. But there is no denying that Freud's 'structures of the mind' framework provides an excellent (additional) means of making sense of how logically incompatible forces can operate simultaneous within the mental apparatus of intrinsically 'rational' human beings.

4.3.2 Solms on Consciousness, the Body, and the Structures of the Mind

In his 2013 article, *The Conscious Id*, Mark Solms described what he considered to be quite a serious flaw in Freud's account of the structures of the mind. However, I propose that a correct understanding of Freud's views on (un)consciousness and (ir)rationality makes it possible to understand Freud in such a way that Solms' arguments do not threaten Freud's framework. I argue that Freud's framework provides an excellent way not only of understanding both conscious and unconscious 'psychical processes' as part of the natural world [Sect. 4.2], but also of accounting for the fact that intrinsically 'rational' persons can be motivated simultaneously by logically incompatible forces [Sect. 4.3].

On Solms' view, the phrase 'the body' has two importantly distinct senses, the 'external body' sense and the 'internal body' sense; and this is crucial to the first respect in which his reading of Freud differs from mine. The 'external body,' for Solms, comprises 'maps,' in the different sensory modalities, of the location and state of different parts of the body: in effect, when we perceive the body in this 'external' sense, we perceive it as just one object among other objects, and it is represented in the brain by the same structures responsible for the sensory processing of any perceptual objects in the environment. The 'internal body' comprises the body's "internal milieu" (2013: 6). The brain structures responsible for this aspect of body representation are responsible not only for 'centralized' sensations like thirst or fatigue, but also for maintaining homeostasis: regulating the overall state of the body in order to maintain healthy life.

These two different senses or aspects of the body, Solms thinks, correspond to two different types of consciousness. The external body relies on 'external (sensory) perception' of the environment and/or of the body itself. Therefore, the external body is associated with perceptual consciousness (or 'easy-problem consciousness,' see Sect. 3.2). The 'internal body,' on the other hand, is "the *subject* of perception … the background state of *being* conscious" (2013: 7)—that is, it is associated with 'consciousness' in the hard-problem sense [compare: Sect. 3.3]. Solms says that "[w]e may picture this aspect of consciousness as the page upon which external perceptions are inscribed" (ibid.).

Then Solms thinks that, having looked at his outline of these two bodily aspects, we will readily perceive that they correspond to "the two major mental systems" that Freud identified:

1. the external body corresponds to the 'ego,' the internal body to the 'id'. (2013: 7)

But according to Solms, looking more closely at 'the internal body' and 'the external body' suggests that, contrary to what Freud said,

2. the id is the fount of consciousness and the ego is unconscious in itself. (2013: 5)

I am going to examine Solms' case in two parts. In this chapter, I will discuss the first two respects in which I consider Solms' claims to be problematic. I will examine the remainder of Solms' case against Freud in Chap. 5, seeing that, in order to evaluate it, I first need to introduce some basic information about brain structures.

4.3.2.1 Psychical Systems vs Ways of Representing the Body

The first thing I am concerned about is Solms' saying that the internal body and the external body correspond to 'the two major mental systems' that Freud recognized, namely the ego and the id. This assumption is crucial to Solms' case—but the question is, *did* Freud actually recognize 'two major mental systems'? Or perhaps more to the point, even if he definitely did recognize 'two major mental systems,' *were* the systems he was thinking of the id and the ego, or were they some other systems?

4.3.2.1.1 Psychical Systems

It is true that, in *An Outline of Psychoanalysis*, Freud, having discussed the id, says that "*[t]he other* agency of the mind" (1940a, in *Complete Works*,

p. 5005, my emphasis) is the ego. This is presumably why Solms assumes that the two major mental systems Freud recognized were the ego and the id. But I am not convinced.

The Unconscious comprises a detailed discussion of conscious and unconscious phenomena. In it, Freud frequently uses the phrase "the system *Cs.* (or *Pcs.*)"—where *Cs.* stands for 'conscious' and *Pcs.* for 'preconscious'—which suggests that he is taking 'conscious' and 'preconscious' as just about interchangeable (compare: Sect. 4.2.1.3; I suggested much the same thing). When he remarks that "processes in the one system, the *Ucs.* ['Unconscious'], show characteristics which are not met with again in the system immediately above it" (1915, in *Complete Works*, p. 3009), it is true that he does not explicitly state which 'the system immediately above' the system *Ucs.* is; but it seems a fairly safe bet that if the first system is "the *Ucs.*," the system immediately above it will be "the *Cs.* (or *Pcs*)." Moreover, as I have already mentioned [Sect. 4.2.2.3], Freud said that the dream-work allows us to study 'what unsuspected psychical processes can occur between *two separate psychical systems like the conscious and unconscious.*' It would therefore probably be safe to conclude that the (i) *Ucs.* and (ii) *Cs.* (or *Pcs.*)—and *not* the (i) Id and (ii) Ego (or superego)—are "the two (or three) psychical systems" (Freud 1915, in *Complete Works*, p. 2997) Freud is referring to.

On Solms' reading, it would be necessary to think of a plausible way of explaining how the three elements of the structural model (id/ego/superego) could somehow be fitted into two 'major mental systems'—Solms' solution seems to be simply to assume that the ego and the id are the only two structures which count for anything, and that the superego is not really one of the structures of the mind. But this is problematic. We know that the ego and the superego are both, according to Freud, only specially modified parts of the id [Sect. 4.3.1.2]—so what makes Solms think that the superego is less worthy of being counted as a major mental system than the ego? One might have expected Freud to say that there was strictly speaking only *one* 'major mental system' (namely, the id), which could, for purposes of convenience, be divided into *three* systems (that is, id/ego/superego)—in fact, on my reading, this is exactly what he did say. On the other hand, it seems almost impossible to think of a good reason for dividing up the tripartite id into *two* systems, of any description whatsoever. It seems much more plausible to suggest that the two major mental systems Freud was referring to are the Conscious (and/or Preconscious), and the Unconscious—and not the ego and the id.

4.3.2.1.2 Ways of Representing the Body

But even if we accept that on Freud's view there were two major mental systems (rather than one, or three), and that these systems were the id and the ego (as opposed to the *Ucs.* and the *Cs./Pcs.*), we still need to consider whether these two systems correspond with Solms' two senses of internal and external body.

Solms is suggesting that 'the external body'—the body as it is perceived as just one object among many—corresponds with the ego. But this is problematic. The function of the ego, according to Freud, is supposed to be perceiving external reality, and reconciling that reality with the demands *of the id*—that is, with the sensations emanating from the interior of the body. But if the ego is concerned with reconciling the demands of the id with the constraints of reality, then it must be able to perceive both the demands of the id and the external reality with which these demands have to be reconciled. And if the ego has the function of *perceiving* both external reality and the demands of the id, it is difficult to see how the ego could itself be taken to correspond to 'the external body' (which is itself a part of that same external reality the ego is tasked with perceiving). If the ego really did correspond with Solms' 'external body,' one would be obliged to say that the function of *the external sense (or aspect) of body representation* [Sect. 4.3.2] was to reconcile the body's instinctual drives with the constraints of reality—and it is nonsensical to suggest that any kind of 'representation' could play such a functional role. 'Representations' are just not the kinds of things that are capable either or *perceiving* things, or of *reconciling* things with other things (which is what the ego does).

That seems like a good reason for thinking that Solms' 'external/internal body' do not correspond to Freud's ego/id. Therefore, seeing that Freud's two major mental systems do not correspond to Solms' internal body and external body, Solms' claim—that (i) 'The external body corresponds to the ego, the internal body to the id'—is problematic.

4.3.2.2 The Structures of the Mind vs the Structure of Consciousness

The second thing I am concerned about is Solms' assumption that Freud took the ego to be conscious and the id to be unconscious. This assumption is crucial to Solms' case against Freud's view—which means that we need to ask: *Did* Freud actually say that the ego was conscious and the id was unconscious?

4.3.2.2.1 Consciousness and the Ego

Freud certainly did say that the ego operated in accordance with the reality principle, and that its main job was (consciously) to perceive reality, and reconcile the constraints of reality with the demands of the id; but the phenomenon of repression suggests that the ego also contains some dynamically unconscious contents:

> [W]e find during analysis that, when we put certain tasks before the patient, he gets into difficulties ... We then tell him that he is dominated by a resistance; but he is quite unaware of the fact, and, even if he guesses from his unpleasurable feelings that a resistance is now at work in him, he does not know what it is nor how to describe it. Since, however, there can be no question but that this resistance emanates *from his ego* and belongs to it, we find ... something in the ego itself which is also unconscious, which behaves exactly like the repressed—that is, which produces powerful effects without itself becoming conscious and which requires special work before it can be made conscious. (Freud 1923b, in *Complete Works*, p. 3951 (my emphasis).[8]

"The ego, in other words, also has an unconscious aspect" (Taylor 2009: 12)—and because this unconscious material in the ego cannot become conscious without that special (psychoanalytic) work, we are not even able to say that it is innocuously unconscious in the sense of latent/preconscious. It is unconscious in the true, dynamic sense. The ego thus proves to be *both* the agency most closely connected to ordinary 'consciousness' *and* the agency through which repression (that is, the process of *removing* something from 'consciousness') takes place. For this reason, it became necessary for Freud to postulate "a third *Ucs.*" (Freud 1923b, in *Complete Works*, p. 3952)—a 'temporary' dynamic unconscious, which is part of the ego—since this seems like the only plausible way of accounting for the phenomenon of repression.

This being the case: it seems that, on the mature 'structural' model of the mind, the topographical picture of *Ucs./Pcs./Cs.* could almost be said to have fallen away—at the very least, the discovery of dynamically unconscious content in the ego "makes the division of the psychic architecture into systems *Ucs* and *Cs* (*Pcs*) less compelling" (Lapsley and Stey 2011: 4). Freud said "[w]e can now play about comfortably with our three terms, *Cs.*, *Pcs.*, and *Ucs.*, so long as we do not forget that in the descriptive sense there are two kinds of unconscious [that is, *Pcs.* and *Ucs.*], but in the dynamic sense only one [that is, *Ucs.*]" (Freud 1923b, in *Complete Works*, p. 3950). Therefore, it seems plausible to suggest not only that Solms may not be justified in ascribing to Freud the view that the ego is conscious, but also that, on Freud's mature view, the question of whether the ego is conscious or unconscious is a question of little significance.

4.3.2.2.2 Consciousness and the Id

The other side of this coin is that Solms also assumes Freud took "unmonitored instinctual processes" in the id as necessarily unconscious (Solms 2013: 11); but I am going to deny this as well. Freud certainly did claim that id processes were deeply and permanently 'unconscious,' but I suggest that psychic material can easily be deeply and permanently 'unconscious' in the perceptual (and/or introspective) sense at the same time as *not* being 'unconscious' in a rudimentary affective sense (that is, the sense in which basic feelings or emotions are 'conscious': they *feel like* something).

Freud himself seems to have distinguished between perceptual consciousness and affective consciousness. As Solms notes, he said that

> What consciousness yields consists essentially of perceptions of excitations coming from the external world and of feelings of pleasure and unpleasure which can only arise from within the mental apparatus. (Freud 1920, p. 24, in Solms 2013: 9)

Now, when Solms uses this quote (along with more paragraphs which I did not include here), he is concerned with demonstrating that Freud assumed that consciousness was produced by the brain's cerebral cortex (more on that in Chap. 5)—but I am going to use it against Solms, with regard to his claim that the id is conscious, and that the ego is unconscious, as follows:

In the first place, 'the perceptions of excitations coming from the external world' (that is, perceptual consciousness), can be safely ascribed to the ego. We know that the ego's job is to perceive reality, and reconcile it, as far as possible, with the desires of the id. In the second place, 'the feelings of pleasure and unpleasure,' arising from within the mental apparatus, are the domain of the id. Freud did say that the id was deeply and permanently 'unconscious'—but if it is to have any kind of commerce with feelings of pleasure and unpleasure, then it must be able, in at least some sense, to be 'conscious' of those feelings. If it were not 'conscious,' in at least some basic sense, of how something felt, the id could not possibly said to experience pleasure and unpleasure, or have those 'urges' that the ego was responsible for reconciling with reality. So it seems quite reasonable to suggest that, in Freud's framework, we have on the one hand the ego, which is (introspectively) 'conscious' of perceptions, and various reality-related considerations; and on the other, we have the id, which is *un*conscious in the perceptual and/or introspective sense, but '*conscious*' of the (affective) feelings produced by biological or instinctual drives (for example, Panksepp 2012: 13).

According to Breuer, Freud, and even the hypnotist Charcot, hysterical symptoms result when a mental process, strongly laden with affective feelings, is repressed. During psychoanalysis, the patient is able to become conscious of the repressed content which was previously inaccessible, and to experience the associated feelings: this is what effects the cure (for example, Freud 1923a, in *Complete Works*, p. 3913). But it was never those actual feelings themselves which were unconscious. Strictly speaking, said Freud, "there are no unconscious affects as there are unconscious ideas" (1915, in *Complete Works*, p. 3001). Whereas "[a]n idea remains the same … whether it is conscious or unconscious … an affect is a process of discharge and must be judged quite differently from an idea" (1916–1917, in *Complete Works*, p. 3455). In the case of ideas, we can accurately say that the ideas are repressed, insofar as they are unavailable to introspection. But "[a]n affective or emotional impulse is *necessarily felt*" (Redding 2000: 124). It makes no sense to call something a feeling (or emotion) *if it does not feel like anything* (compare: Solms 2014: 173)!

Being 'conscious' of instincts or feelings is such a limited sense of being conscious, compared with being *introspectively* conscious (of perceptions, cognitions, thoughts, propositions et cetera), that I think it would still be in the spirit of Freud's insights to suggest that Freud could have used the terms 'conscious' and 'unconscious'—with reference to the ego and the id—in the sense of 'conscious-of-(external) perceptions,' and '(only)-conscious-of-(internal) subjective-feelings,' respectively. The id is 'deeply and permanently *un*conscious' in the sense that the instinctual, unconscious drives it produces are not available to introspection under any circumstances; but it *is* 'conscious' in the sense that the individual is 'aware' of the affective sensations of pleasure and unpleasure which arise from these unconscious drives.

This being the case, Solms' contention, that his 'discovery'—that (2) 'the id is the fount of consciousness and the ego is unconscious in itself'—presents a challenge to Freud's framework, is problematic. I have suggested that Freud viewed the id as unconscious in the perceptual and/or introspective sense, but *conscious* in the very limited sense of being aware of its own feelings. In this case, that very claim which Solms takes to be problematic for Freud's framework (that the id is 'conscious') might be exactly the claim that Freud himself was making (that the id is not *introspectively* conscious/is conscious only in an affective sense).

4.3.2.3 The Lesson (of 'Solms on Consciousness, the Body, and the Structures of the Mind') for Philosophy

The apparent difficulty about whether to say that the id/ego are un/conscious therefore turns out to be a 'verbal dispute' only [compare: Sect. 4.2.1.3.1]. If one is thinking of 'consciousness' in the perceptual and/or introspective sense, then the instinctual processes in the id are trivially, by definition, *not* 'conscious.' And if one is thinking of 'consciousness' in the affective sense, then the instinctual (affective) processes of the id *are* by definition (but certainly not 'trivially') 'conscious.' On my reading, so far from having "massive implications for our conceptualization of the ego and all that flows from it" (Solms 2013: 12), Solms' idea of 'the conscious id' seems to fit rather well into Freud's conceptual framework, as I have set it out.

4.3.3 (Psychological) What-It-Is-Like/Experience and Rationality: Conclusion

In Sect. 4.3, I have shown that the Freudian view of 'irrationality' is considerably more satisfactory than the 'accounts' of irrationality which philosophers tend to produce—which might more accurately be described as 'non-accounts.'

While it is not (strictly speaking) possible to give a coherent account of 'irrationality,' both Freud's work and more recent studies suggest that psychiatric symptoms result when patients' consciously-held values are logically incompatible with unconscious psychological material which has been imperfectly repressed. Freud's 'structural model' is particularly effective in enabling him to explain how conflicting motives could operate simultaneously within a single individual, without undermining the assumption that we are intrinsically rational. Moreover, on the psychodynamic view, the most important part of the psychic apparatus—in the dynamic, economic, topographic, or structural sense—is not even close to being 'rational.'

Even if Solms correctly identified Freud's 'two major mental systems' *and* correctly interpreted Freud's views about whether these systems were or were not 'conscious,' it is possible to say without contradiction that the id is 'unconscious' in the perceptual (and/or introspective) sense, at the same time as being 'conscious' in a rudimentary affective sense. Similarly, it is possible to say that the ego is essentially conscious in a perceptual (and/or introspective) sense, while at the same time containing some dynamically unconscious content. On my reading, so far from having 'serious implications' for psychodynamic metapsychology, Solms' 'discovery' of 'the conscious id' is more or less what Freud meant all along, and there is nothing in any of Solms' claims to prevent us from taking a Freudian-type view of a psychological subject which is not (essentially conscious or) essentially rational.

Because, on the psychodynamic account, some of the most influential factors contributing to both psychological and somatic experience are logically incompatible with one another—and because the psychodynamic account seems to be (virtually) the only satisfactory account of irrationality we have—we should be careful not to conflate (psychological) what-it-is-like/experience with 'rationality.'

4.4 (Un)Consciousness and (Ir)Rationality in Psychology: Conclusion

So, I have argued that we cannot talk about 'the explanatory gap' (and/or 'the mind' and/or 'consciousness') in a literal sense unless we are dualists [Sects. 2.2 and 2.3]; I have explained that, because both perception and cognition can be, and often are, 'unconscious,' we should be careful not to conflate 'consciousness' with (the easy problems of) 'conscious' perception and cognition [Sect. 3.2]; and that, because there is nothing-it-is-like to experience unconscious (and/or unattended) perceptual or cognitive processes, we should also be careful not to conflate (hard problem) what-it-is-like/experience with (perceptual-cognitive) 'consciousness' [Sect. 3.3].

In addition, in this chapter, I have now shown that:

1. Because the existence and significance of dynamically unconscious material can be both theoretically accounted for and empirically corroborated, we should be careful not to conflate (psychological) what-it-is-like/experience with (introspective) 'consciousness' [Sect. 4.2].
2. Because so many psychodynamic processes are 'irrational'—and because the psychodynamic account seems to be (virtually) the only satisfactory account of irrationality we have—we should be careful not to conflate (psychological) what-it-is-like/experience with 'rationality' [Sect. 4.3].

In fact, trying to understand what-it-is-like/experience in terms of (introspective) 'consciousness' or rationality (see this chapter) is just as problematic as trying to understand either 'consciousness' or what-it-is-like/experience in terms of perception and/or cognition (Chap. 3)—all of which, moreover, are just as problematic as trying to understand left- and right-hand gloves in terms of pairhood (Chap. 2).

The Mind-Body-Self:

This being the case, there are a number of our customary ways of speaking which are no longer available to us. I have explained why we need to stop using terms like 'the mind' (so long as we are not dualists [Sect. 2.2]); or 'consciousness' (so long as we recognize that a large proportion of both perception/cognition [Sect. 3.2], and 'psychological material' [Sect. 4.2], is not available to conscious introspection). Moreover, I have explained why it appears to be inadvisable to use the term 'rationality' when referring to anything to do with (psychological) what-it-is-like/experience [Sect. 4.3]. In fact, it is difficult to know how to talk about what we want to talk about, without committing some quite severe oxymorons.

Writers like Panksepp (2012) and Davis and Panksepp (2011) use terms like 'mind-brain' and 'BrainMind,' because they are committed to the rejection of mind-body dualism. But I think I need to advance this trend even further. I am not only interested in the metaphysical aspect of the problem of 'consciousness,' but also in the psychological subject of experience. Therefore, from this point on, I will make use of the term 'mind-body-self,' to indicate a *very* broad and inclusive understanding of what we have formerly been accustomed to thinking of as 'the mind' and/

or 'the body' and/or 'the subject of (psychological) what-it-is-like/experience.' By making use of this term, I mean to signify explicitly that I am not committing myself to any definite metaphysical position on any of these questions (except of course insofar as I am committing myself to the claim that dualism is not acceptable, since I am committed to 'biological naturalism').

Accordingly, if we truly want to make any progress in our investigation of *the mind-body-self*, we need to abandon the futile debate about whether dualism or materialism is preferable (Chap. 2); we need to abandon the kinds of investigations which are focused on perceptual (or 'sensory') or cognitive awareness (Chap. 3); and we need to abandon the kinds of investigations which focus on (introspective) consciousness or rationality (this chapter). We need to focus—as Freud does—on the psychological and biological, predominantly unconscious, predominantly irrational, mind-body-self. In order to do that, we need to find out more about how 'the psychical apparatus' works—and the literal 'psychical apparatus' (as even Descartes knew perfectly well, let alone, as Freud knew perfectly well) is the brain. This 'literal psychical apparatus' will be the topic of Chap. 5.

Notes

1. Freud himself tends not to use terms like 'the mind' or 'conscious*ness*'—I will argue [Sect. 4.2.1.1] that this is because he assumes, as I do, that dualism is not plausible.
2. Strictly speaking, 'psychodynamic theory' is a theory of *personality*; but I am only interested in the elements of this theory that inform the (psychodynamic) metaphysical view of 'the mental apparatus.'
3. Even here, in one of Freud's clearest statements of his commitments to physicalism, it seems that Wakefield (2018: 253) manages to misunderstand entirely: he says "The basic message in [this quotation] is that it is scientifically disastrous to require psychologists to grapple with the insoluble problem of consciousness"—whereas the actual basic message, I would say, is more like 'Grappling with the problem of consciousness requires dualism, which is an untenable position.'

4. Whether or not something similar to this vesicle actually exists is a moot point: in giving this account of the primitive vesicle, Freud was making a logical claim about what capacities organisms would need to have, rather than an empirical claim about what organisms are like. We were directed to *imagine* a living organism in its most simplified form, not to *call to mind* an organism with which we were familiar. For Freud, 'living vesicle' is a vague term that entails no particular ontological commitments—perhaps similar to 'elementary particle.'
5. It is difficult to see why Boag (2015: 253) thinks that the quoted phrase is an attempt on Freud's part "to capture the relationship between knowing and knowing that one knows": Freud could not have been much more explicit than he is here about what exactly he is trying to do, which is to show why psychoanalysis cannot accept the conventional equation of 'psychical' with 'conscious.' He is reminding us that, since it is uncontroversial to claim that everybody always has a whole lot of memories and 'knowledge' that are not 'conscious,' people already implicitly acknowledges the existence of 'the unconscious.'
6. Just as, in this chapter, I have chosen to focus in depth on psychodynamic theory (rather than focusing more superficially on a number of theories), so I have chosen here to focus in depth on the criticism of Freud presented by Solms, rather than focusing in less depth on more criticisms.
7. Note that Freud's view concerns 'the unconscious' (as a 'province' within the psychical apparatus) and *not* (literally) 'unconscious*ness*.' Unconsciousness is the state somebody is in after (for example) sustaining a head injury.
8. See also 1920, in *Complete Works*, p. 3724.

References

Boag, S. 2015. In Defence [sic] of Unconscious Mentality. In *Psychoanalysis & Philosophy of Mind: Unconscious Mentality in the 21st Century*, ed. S. Boag, L.A.W. Brakel, and V. Talvitie, 239–265. ProQuest Ebook Central database: https://0-ebookcentral-proquest-com.ujlink.uj.ac.za/lib/ujlink-ebooks/detail.action?docID=2082961&query=psychoanalysis+and+philosophy+of+mind. Accessed 10 September 2019.

Chalmers, D.J. 1996. *The Conscious Mind: In Search of a Fundamental Theory*. https://pdfs.semanticscholar.org/5491/3945b27ca973968174ad521aa8562 0f0b32f.pdf. Accessed 7 August 2016.

Churchland, P.S. (1986). Neurophilosophy: Toward a Unified Science of the Mind-brain. Available from: https://www.federaljack.com/ebooks/Consciousness%20Books%20Collection/Patricia%20Churchland%20-%20Neurophilosophy.pdf

Davis, K.L., and J. Panksepp. 2011. The Brain's Emotional Foundations of Human Personality and the Affective Neuroscience Personality Scales. *Neuroscience and Biobehavioral Reviews* 35: 1946–1958.

Ellenberger, H.F. 1972. The Story of "Anna O": A Critical Review with New Data. *Journal of the History of the Behavioral Sciences* 8 (3): 267–279. https://doi.org/10.1002/1520-6696(197207)8:3<267::AID-JHBS2300080302>3.0.CO;2-C.

Freud, S. 1920. Beyond the Pleasure Principle. In *The Standard Edition of the Complete Works of Sigmund Freud* (1955), trans. James Strachey in collaboration with Anna Freud. London: The Hogarth Press and the Institute of Psycho-Analysis.

Freud – Complete Works, Ivan Smith 2000, 2007, 2010. https://www.valas.fr/IMG/pdf/Freud_Complete_Works.pdf. Accessed 20 September 2015.

———. 1910. Five Lectures on Psycho-Analysis. In *Complete Works*, pp. 2195–2239.

———. 1912. A Note on the Unconscious in Psycho-Analysis. In *Complete Works*, pp. 2575–2583.

———. 1913. On Psycho-Analysis. In *Complete Works*, pp. 2543–2549.

———. 1915. The Unconscious. In *Complete Works*, pp. 2989–3024.

———. 1916–1917. Introductory Lectures on Psycho-Analysis. In *Complete Works*, pp. 3124–3501.

———. 1920. Beyond the Pleasure Principle. In *Complete Works*, pp. 3713–3762.

———. 1923a. Two Encyclopaedia Articles. In *Complete Works*, pp. 3910–3934.

———. 1923b. The Ego and the Id. In *Complete Works*, pp. 3944–3992.

———. 1924. A Short Account of Psycho-Analysis. In *Complete Works*, pp. 4099–4118.

———. 1926. Psycho-Analysis. in *Complete Works*, pp. 4399–4407.

———. 1940a. An Outline of Psycho-Analysis. In *Complete Works*, pp. 4954–5011.

———. 1940b. Some Elementary Lessons in Psycho-Analysis. In *Complete Works*, pp. 5065–5071.

Gallagher, S., and D. Zahavi. 2008. *The Phenomenological Mind: An Introduction to Philosophy of Mind and Cognitive Science*. https://doi.org/10.4324/9780203086599.

Gay, P. 1989. Introduction to Civilization and Its Discontents. In *The Standard Edition of the Complete Psychological Works of Sigmund Freud*, ed. J. Strachey, 2nd ed., 5–9. New York: W.W. Norton.

Hopkins, J. 2015. The Significance of Consilience: Psychoanalysis, Attachment, Neuroscience, and Evolution. In *Psychoanalysis & Philosophy of Mind: Unconscious Mentality in the 21st Century*, ed. S. Boag, L.A.W. Brakel, and V. Talvitie, 47–136. ProQuest Ebook Central database: https://0-ebookcentral-proquest-com.ujlink.uj.ac.za/lib/ujlink-ebooks/detail.action?docID=2082961&query=psychoanalysis+and+philosophy+of+mind.

Kandel, E.R. (1999). Biology and the Future of Psychoanalysis: A New Intellectual Framework for Psychiatry Revisited. *American Journal of Psychiatry*, 156: 505–524. Available from: https://ajp.psychiatryonline.org/doi/pdf/10.1176/ajp.156.4.505

Kaplan, R. 2004. O Anna: Being Bertha Pappenheim – Historiography and Biography. *Australasian Psychiatry* 12 (1): 62–68.

Lapsley, D.K., and P.C. Stey 2011. Id, Ego, and Superego. In *Encyclopedia of Human Behavior*, 2nd ed. https://www3.nd.edu/~dlapsle1/Lab/Articles%20&%20Chapters_files/Entry%20for%20Encyclopedia%20of%20Human%20Behavior%28finalized4%20Formatted%29.pdf. Accessed 4 April 2018.

Levine, M.P. 2000. Introduction: How Right Does Psychoanalysis Have to Be? In *The Analytic Freud: Philosophy and Psychoanalysis*, ed. Michael Levine, 1–7. London: Routledge.

Marshall, G. 2000. How Far Down Does the Will Go? In *The Analytic Freud: Philosophy and Psychoanalysis*, ed. Michael Levine, 36–48. London: Routledge.

Middleton, W. 2004. Anna O and Hanna Q. *Australasian Psychiatry* 12 (3): 239–244.

Panksepp, J. 1998. *Affective Neuroscience: The Foundations of Human and Animal Emotions*. ProQuest Ebook Central database: https://0-ebookcentral-proquest-com.ujlink.uj.ac.za/lib/ujlink-ebooks/detail.action?docID=431176#.

———. 2004. Biological Psychiatry Sketched – Past, Present, and Future. In *Textbook of Biological Psychiatry*, J. Panksepp, 3–32. ProQuest Ebook Central database: https://0-ebookcentral-proquest-com.ujlink.uj.ac.za/lib/ujlink-ebooks/reader.action?docID=469784.

———. 2012. A Synopsis of Affective Neuroscience – Naturalizing the Mammalian Mind. *Journal of Consciousness Studies* 19 (3–4): 6–18.

Pataki, T. 2015. Wish-Fulfilment Revisited. In *Psychoanalysis & Philosophy of Mind: Unconscious Mentality in the 21st Century*, ed. S. Boag, L.A.W. Brakel, and V. Talvitie, 1–46. ProQuest Ebook Central database: https://0-

ebookcentral-proquest-com.ujlink.uj.ac.za/lib/ujlink-ebooks/detail.action?docID=2082961&query=psychoanalysis+and+philosophy+of+mind.

Redding, P. 2000. Freud's Theory of Consciousness. In *The Analytic Freud: Philosophy and Psychoanalysis*, ed. Michael Levine, 101–118. London: Routledge.

Ryle, G. 1949. *The Concept of Mind*. London: Routledge.

Snelling, D. 2000. Some Reflections on the Kleinian Origins of the Mind. In *The Analytic Freud: Philosophy and Psychoanalysis*, ed. Michael Levine, 101–118. London: Routledge.

Solms, M. & Panksepp, J. (2012). The "id" Knows More than the "Ego" Admits: Neuropsychoanalytic and Primal Consciousness perspectives on the interface Between Affective and Cognitive Neuroscience. *Brain Sciences* 2: 147–175.

Solms, M. 2013. The Conscious Id. *Neuropsychoanalysis: An Interdisciplinary Journal for Psychoanalysis and the Neurosciences* 15 (1): 5–19. https://doi.org/10.1080/15294145.2013.10773711.

———. 2014. A Neuropsychoanalytic Approach to the Hard Problem of Consciousness. *Journal of Integrative Neuroscience* 13 (2): 173–185.

Sulloway, F.J. 1979. *Freud, Biologist of the Mind: Beyond the Psychoanalytic Legend*. Suffolk: The Chaucer Press.

Talvitie, V. 2015. Beyond the Philosophy of the (Unconscious) Mind: The Freudian Cornerstone as Scientific Theory, a Cult, and a Way of Talking. In *Psychoanalysis & Philosophy of Mind: Unconscious Mentality in the 21st Century*, ed. S. Boag, L.A.W. Brakel, and V. Talvitie, 163–192. ProQuest Ebook Central database: https://0-ebookcentral-proquest-com.ujlink.uj.ac.za/lib/ujlink-ebooks/detail.action?docID=2082961&query=psychoanalysis+and+philosophy+of+mind.

Tanney, J. 1949. Rethinking Ryle: A Critical Discussion of *The Concept of Mind*. In *The Concept of Mind*, ed. G. Ryle, ix–lvii. London: Routledge.

Taylor, E. 2009. *The Mystery of Personality: A History of Psychodynamic Theories*. New York: Springer.

Wakefield, J.C. 2018. *Freud and Philosophy of Mind, Volume 1: Reconstructing the Argument for Unconscious Mental States*. https://doi.org/10.1007/978-3-319-96343-3.

Westen, D. 1999. The Scientific Status of Unconscious Processes: Is Freud Really Dead? *Journal of the American Psychoanalytic Association* 47 (4): 1061–1106. https://doi.org/10.1177/000306519904700404.

5

The Brain and the Mind-body-self

I explained [Sect. 4.1] that the main reason behaviorism was ever popular (in psychology, especially America) was that it was apparently so 'scientific'—and even now, when the main behaviorist heyday has largely come to an end, the behaviorist legacy lives on in virtually all psychological theory and practice. But the behaviorists had no interest in the actual mechanisms of stimulus, reinforcement and behavior: as long as they could observe *that* certain stimuli reliably led to certain behaviors, they were uninterested in *how* or *why* these processes actually worked.

The crucial difference between psychodynamics and behaviorism is that, whereas behaviorism stipulated that *as a matter of principle* there was nothing interesting to be said about the physical (neural) mechanisms underlying its behavioral principles of stimulus and reinforcement, Freud always worked on the assumption that the physical account should be regarded as more complete than the psychological one [Sect. 4.2.1.1]. He thought that 'the deficiencies in our description [of the mental apparatus] would probably vanish if we were already in a position to replace the psychological terms by physiological or chemical ones' [Sect. 4.2.1.2].

But because of the limitations of Freudian-era neuroscience, Freud ultimately had no choice but to abandon work on his drafted *Project for a Scientific Psychology* (1895)—more on that in Sect. 5.1.1.1—and

© The Author(s), under exclusive license to Springer Nature Switzerland AG 2022
J. Anderson, *Biological Naturalism and the Mind-Body Problem*,
https://doi.org/10.1007/978-3-030-99684-0_5

continue to try and express his ideas using the 'figurative language peculiar to psychology' [Sect. 4.2.1.2]. Although this "permitted the development of systematic definitions of behavior and disease that were not contingent on still-vague correlations with neural mechanisms" (Kandel 2005: 35), Freud ultimately failed to achieve his goal of producing a naturalistic account of (psychological) what-it-is-like/experience, precisely because these definitions were not securely grounded on a good understanding of brain function.

In the twenty-first century, the idea that the brain is 'the bodily organ of mental life' [compare: Sect. 4.2.1.1] "is so central in traditional thinking in biology and medicine … that it is almost a truism and hardly needs restatement" (Kandel 2005: 40). But although this truism is generally accepted, precisely how it is that mental life comes from a bodily organ is still poorly understood. Even if the more enlightened of the twentieth-century mind science researchers did not literally believe that 'the mind' and the brain were separate, a serious outstanding problem was "they did not know how to join them" (Kandel 2005: xxii). Now that we know a lot more about the brain than even the best physicians did in Freud's day, or even in most of the twentieth century, Kandel suggests that using new techniques to investigate Freudian ideas empirically is just what we need to help us develop "a new and compelling perspective on the mind and its disorders" (2005: 65).

Such a perspective, encompassing principles of both (neuro)biology and psychodynamics, would enable clinicians to make sense of psychological problems (and develop effective treatments for them), while satisfying philosophers with its cogent internal logic. In fact, it sounds as if this might be exactly the kind of framework I have been looking for, within which to account for thinking/consciousness (or what-it-is-like/experience) in terms of the functioning of physical structures.

5.1 Introduction: The Cerebral Cortex

In botany, the word 'cortex' can refer to any kind of 'outer layer,' like the bark of a tree. With regard to living creatures, it refers to the outer region of an organ or structure, but particularly the outer layer of the hemispheres of the brain, which is the *cerebral* cortex. The cerebral cortex consists of large sheets of neurons covering over the deeper structures of the brain. In humans and higher mammals, the cortical layer is not stretched smoothly over these deeper brain structures, but highly folded and convoluted—with the result that a much higher (cortical) surface area can be contained in a much smaller space within the skull. The brain is divided into the frontal lobe, the parietal lobe, the temporal lobe, and the occipital lobe; and the cerebral cortex extends over all four of these lobes.

The somatosensory (cerebral) cortex (in the *parietal* lobe) receives sensory input from all sensory modalities—including the sense of touch, which incorporates the feeling of pain, temperature sense, and the sense of where the limbs are positioned in space. The primary visual cortex is in the *occipital* lobe: this is where almost all visual processing takes place (I already mentioned that this is where the damage to the brains of blindsight patients is found, Sect. 3.2.1). The cortex of the *temporal* lobe is largely devoted to auditory processing. In the *frontal* lobe of the brain, the cortex comprises the prefrontal cortex (which I will discuss in detail below), in the front; and the motor cortex, which is quite a narrow strip that goes over the top of the head, from ear to ear. The motor cortex contains motor neurons which connect with the spinal cord, and is essential for all types of bodily movement. Finally, some of the cortex is devoted neither to sensory processing nor to motor function (ibid.); this is known as the '*association* cortex,' since these cortical regions receive inputs from different sensory modalities, which they then have to integrate. In humans and higher mammals, there is more association cortex in proportion to overall brain size than there is in other animals.

As I said [Sect. 3.1], twentieth-century mind-science researchers have tended to focus on perceptual and cognitive types of 'consciousness.' And since almost all the brain functioning related to our five senses, as well as the functioning related to most cognitive processing, goes on in the cerebral cortex, this also means that these researchers have been looking almost exclusively at *cortical* types of consciousness (for example, Baars 2005a: 45, b: 7). From now on, I'll be taking this for granted: insofar as a certain view centres around perception and/or cognition, that view is (also, therefore) 'corticocentric.'

I have now determined that we should not conflate 'consciousness' with ('conscious') perception or cognition [Sect. 3.2]; we should not conflate what-it-is-like/experience with (perceptual-cognitive) 'consciousness' [Sect. 3.3]; we should not conflate (psychological) what-it-is-like/experience with (introspective) consciousness [Sect. 4.2]; and we should also not conflate (psychological) what-it-is-like/experience with rationality [Sect. 4.3]. In Chap. 5, I am going to examine a neurological condition, called hydranencephaly, in some depth. This will allow me, firstly, to demonstrate that, because hydranencephalic children undoubtedly enjoy some level of what-it-is-like/experience, we should not take the mind-body-self to be dependent on the functioning of the cerebral cortex (which these children lack) [Sect. 5.2].

My discussion of hydranencephaly will also enable me to make my second point, and my first positive claim: because rudimentary affective awareness can be present when perceptual-cognitive and/or introspective-rational 'consciousness' is absent—can be present, in fact, even in individuals (or species) which have no capacity at all for these types of 'consciousness'—we should understand perceptual-cognitive/introspective-rational consciousness as derivative, and rudimentary affective awareness as foundational [Sect. 5.3]. We *should* understand the (subcortical) mind-body-self—which is not essentially perceptual, cognitive, (introspectively) conscious, or rational—as essentially affective.

In the name of clarity, I usually distinguish quite sharply between the 'cortical' and the 'subcortical' parts of the brain—but one should always bear in mind that, even if certain brain structures share the characteristic of being 'cortical' or 'subcortical,' their having this quality in common does not even nearly amount to their being identical, or even necessarily having anything much else in common. Unlike calculators, or even computers, the brain is *not* a straightforward mechanical tool, with a finite number of independent components, functioning in a predictable way.

Moreover, although some generalizations can be made about 'the cerebral cortex' and 'the subcortical brain,' in reality, both the cortical and the subcortical brain are made up of many different structures, intricately interconnected and far from identical. This means that I am to some extent exaggerating the cortical/subcortical opposition in the name of bringing out more clearly the contrast between the different types of 'consciousness' I am interested in.

5.1.1 Historical Views

5.1.1.1 Late Nineteenth- and Early Twentieth-Century Views

5.1.1.1.1 Psychophysical Parallelism

Because it was only when he was in his mid-forties that he finally abandoned biological research, Freud was (at least in the early parts of his career) "steeped in the neurological conceptions of the mind and consciousness of the late nineteenth century" (Redding 2000: 120). And in the late nineteenth century, the prevailing view was 'psychophysical parallelism'—not quite as bad as Cartesian dualism, but a dualistic kind of view nevertheless.[1] For instance, the psychophysical parallelist Gustav Fechner, who seems to have had quite a significant impact on Freud's developing views, came up with the idea of 'the mind' being related to 'the body' in the same way that the concave and convex aspects of a single curved line are related. These two 'aspects' of the line are demonstrably not 'independent' of each other, as Descartes' thinking things and physical bodies were independent—but the 'relation' between them still seems to be one whose nature is somewhat perplexing and hard to explain [Sect. 3.2].

5.1.1.1.2 The Cerebral Reflex

Nineteenth century advocates of psychophysical parallelism thought that the sensory surfaces of the body were 'mapped' "in a topographical point by point manner" (Redding 2000: 122) onto the sensory regions of the cerebral cortex. According to the parallelist view, psychological experience essentially consisted of these bodily sensations, as well as the 'association of ideas' with these sensations. But John Hughlings Jackson (among others) challenged the parallelist views on cortical representation (expressed by Theodore Meynert). Instead, these people supported the notion of the cerebral reflex.

Proponents of the cerebral reflex theory argued that neural traffic from the body's surfaces to the cortex had to pass through 'intermediary' neurons, each of which controlled reflexes of its own. In this case, the role of the cortex could not be purely the topographical representation of regions of the body. Rather, the cortex was responsible for both the simple, mechanical mapping of the body's surface and the 'higher order' mapping *of* the maps of the body which were produced by the 'intermediary neurons.' Therefore, on this view, any scheme (like Meynert's) according to which the cortex is responsible only for the point-by-point mapping of the body's periphery must be seen as problematic.

Freud realized that it would be problematic to try and differentiate sharply between how the brain produced physical (sensory) 'perception' and how it produced the various 'psychological' processes which were associated with these perceptions: he knew that perceptual functions and psychological functions were equally dependent on brain functioning. Just as he knew it was problematic to say that the cerebral cortex was responsible only for mapping the body's periphery, and not the higher order maps *of* these body maps, Freud also knew that to say the brain produced only sensory perception, and not psychological processes, was similarly problematic. Thus the cerebral reflex doctrine was responsible for eventually undermining Freud's commitment to parallelism (Redding 2000: 121).

Freud also realized that "[w]e cannot have a perception without immediately associating it [with various ideas/other psychological content]; however sharply we may separate the two concepts, in reality they belong to one single process which, starting from one point, spreads over the whole cortex (Freud 1953: 57)" (Redding 2000: 122–123). From this point forward, having abandoned his parallelist view, Freud adopted a view about how the brain produces psychological processes were which was essentially the same as (albeit much less detailed than) contemporary views.

Freud started work on the his *Project for a Scientific Psychology* in 1895; although it was never completed, it is right at the beginning that "Freud announces his abandonment of substance dualism in favour of materialism" (Smith 1999: 415) when he says "[t]he intention of this project is to furnish us with a psychology which shall be a natural science: its aim, that is, to represent psychical processes as quantitatively determined states of specifiable material particles and so to make them plain and void of contradictions" (Freud 1895: 355).

> Although nowadays almost all cognitive scientists and philosophers of mind are materialists, this was a very radical step to take in 1895 when dualism held the field. (Smith 1999: 416)

—almost as radical as to suggest listening to the views of one's patients, rather than suspending them from the ceiling [Sect. 4.2.2.1].

5.1.1.2 Twenty-First-Century Views

5.1.1.2.1 The Pre-frontal Cortex

In humans, as I said [Sect. 5.1], the cerebral cortex is more extensive than in other mammal species; but actually, most of the intellectual advantage we have over other mammals can be ascribed to the development of the *prefrontal* cortical area in particular. Because it has communication links with so many other areas of the brain, the prefrontal cortex "is in an excellent position to coordinate processing across wide regions of the central nervous system." The cortical aspects of all four lobes of the brain project to the prefrontal cortex; some subcortical structures also do so; and "[t]he pfc also sends reciprocal connections to most areas that project to it" (Gazzaniga et al. 2002: 501).

Grouped together, the specialized functions of the prefrontal (cortical) region are often referred to as 'executive function': those advanced capacities useful for formulating a plan of action which not only draws on past experience, but is also uniquely suited to a particular current context. A bit further on in the chapter [Sect. 5.3.1], I'm going to discuss the cubomedusa: this is a type of predatory jellyfish whose rudimentary nervous system allows it to move, in response to sensory input, in one of four basic directions. In contrast, we humans are capable of very sophisticated goal-oriented behavior, which is customized to specific situations. We are capable of "reasoning, planning, and performing mental simulations" (Cacioppo and Berntson 2004: 978)—and it is largely the superior development of the prefrontal region of the cortex (and its 'executive function' capabilities) that can account for humans' superior skills in these respects.

> The best route from the dorm room to the eatery might be blocked by a construction project, but a highly evolved system like a hungry college student will still succeed in getting to dinner. To switch from one plan to another requires a means for monitoring the success of our ongoing actions. These operations are commonly referred to as executive functions. (Gazzaniga et al. 2002: 500)

Recall what I said about Churchland in Sect. 3.1: she was interested in finding out how we learn, what makes us smart, how we plan and problem-solve, recollect and forget, et cetera. Those kinds of things are all good examples of capacities which are enabled by the advanced development of the (human) prefrontal cortex.

In Pohl's primate studies [Sect. 3.2.1.2], the monkeys first had to learn a rule (for example, 'The food is in the well close to the cylinder'), and then as soon as they had learned the first rule, the contingency was reversed, and they had to learn a new rule (for example, 'The food is *not* in the well close to the cylinder'). The ability not only to learn a rule about how to find food, but also to learn a second rule, and then alternate between the two rules, is also a paradigmatic 'executive function' (and therefore pfc) task.

The studies on 'attribution' (or 'cognitive dissonance,' Sect. 3.2.3.1.1) relied on people's ability to understand the relation between cause and effect: for example, in the study where the subjects listened to sounds which they believed to be their own heartbeats, while they viewed snakes, I said it was important that the subjects could understand the conceptual relation between feeling frightened and having a rapid heartbeat. I said that those studies would not have worked on anybody who lacked the ability to understand this logical connection; and the ability to understand the logic of cause and effect becomes possible as the result of the advanced development of the cerebral cortex—specifically, of the *prefrontal* cerebral cortex.

In Sect. 4.3, I said that behaviorism (of both the philosophical and psychological varieties) rests on the assumption that people essentially behave in rational ways—to the extent that, for some philosophers, the capacity for rationality is a prerequisite for *any* kind of 'experience' whatsoever [Sect. 4.3.1]. Seeing that such views take humans to be essentially rational, they also imply that being human essentially involves the sophisticated human pfc.

5.1.1.2.2 The Subcortical Brain

Early twentieth-century researchers assumed not only that executive function and higher perceptual function were cortical functions, but that sleep and wakefulness were, likewise (since these are the two underlying 'states' that determine whether or not any of the 'higher' functions can occur). Then Moruzzi and Magoun (1949) published some important research which showed that lesions in certain subcortical sites resulted in 'somnolence' or even coma (Merker 2007a: 64). This was "quickly confirmed" by Penfield and Jasper in 1954 (Solms and Panksepp 2012: 163), and the idea that the brainstem is involved in regulating 'cortical state' (that is, sleep/wakefulness) is now widely accepted.

But many researchers still tend to think that "[t]he old [subcortical] brain core handles basic biological regulation down in the basement, while up above the neocortex deliberates with wisdom and subtlety" (Damasio 1994: 128), and they often do not seriously consider the possibility of the subcortical brain's being involved in a range of quite advanced (hard-problem and easy-problem) 'consciousness' phenomena.[2] Therefore—although of course there are some exceptions—most of the research conducted during the recent "upsurge of interest in the nature and organization of consciousness" (Merker 2007a: 64) has not devoted much attention to subcortical mechanisms.

Contributing to the tendency of researchers to focus on the cerebral cortex, to the exclusion of subcortical structures, is the fact that the subcortical brain is actually very difficult to visualize. Modern brain imaging tools (Positron Emission Tomography, Functional Magnetic Resonance Imaging) "are typically not well suited to visualizing … brainstem [that is, 'subcortical'] … regions" (Panksepp 2012: 15–16). In the brainstem, several interacting systems are "incredibly tightly intermeshed" (Panksepp 1998: 304), which adds to the difficulty of visualizing individual systems. Not only are the brainstem structures physically 'intermeshed' with each other: they also interact significantly with a number of cortical structures, which makes it even more difficult to draw any definite conclusions about the exact role being played by the subcortical brain.

Although visualizing the subcortical brain is difficult, preclinical animal research has long been suggesting that a kind of rudimentary affective awareness is mostly preserved when the cerebral cortex is completely removed. Decorticated animals not only appear to be perfectly capable of experiencing emotion, but "are actually more emotional than normal" (Solms and Panksepp 2012: 161). And because most researchers now recognize that "at a genetic and subcortical-brain organizational level, all mammals are strikingly similar" (Panksepp 2003: 5), the findings of these animal studies can, with a reasonable degree of confidence, be taken to be relevant to humans as well. Moreover, studies on (human) neurological patients "indicate that the subcortical areas are of decisive importance in both the generation of affect and primary-process consciousness;" "[r]ecent work with new brain-imaging approaches" also seems to indicate this (Panksepp 2001: 153). But according to Panksepp, there is as yet no "high-quality work on the underlying neural principles of primary-process affective consciousness … on the Anglo-American science scene" (Panksepp 2005a: 35).

5.2 The Cerebral Cortex and Consciousness

Just as Descartes, I argued [Sect. 2.2.1.2], was aware that the brain was essential for thinking, so too was Freud [Sect. 4.2.1.1]; moreover, Freud was also aware that the brain was essential for *perceiving*. In fact, Freud was aware that almost all thinking, perceiving, and other psychical processes were carried out specifically in *the cerebral cortex* of the brain. What he said was that "[t]he whole organization of the brain seems to fall into two central apparatuses of which the cerebral cortex is the younger, while the older one is represented by the ganglia of the forebrain which have still maintained some of their phylogenetically old original functions" (1953: 49–50, in Redding 2000: 126). Most contemporary neuroscience views would agree with this division; moreover, most would recognize the more evolutionarily recent cerebral cortex as essential for most (easy problems of) perception and cognition, as well as for (the hard problem of) 'consciousness' (for example, Merker 2007a; Panksepp 2007; Watt 2007). However, there is a growing tide of opinion according to which this assumption is not justified.

In the first part of this section, I will discuss a condition called *hydranencephaly*, in which children are born with part or all of their cerebral cortex damaged or even entirely absent. This will allow me to demonstrate that the cerebral cortex is essential neither for hard-problem nor easy-problem 'consciousness' [Sect. 5.2.1]. Having examined some hydranencephaly cases, I will then [Sect. 5.2.2] be in a better position to analyse the second part of Solms' case against Freud (the first part of which I discussed in Sect. 4.3.2). I will consider what Solms says about how specific types of consciousness are produced in the brain, how he thinks the case of hydranencephaly counts against Freud's view, and how I think I am able to undermine Solms' challenge to the Freudian framework.

5.2.1 'Consciousness' in Hydranencephaly

Anencephaly is the term for any case in which the cerebral hemispheres of the brain either fail to develop properly in the first place, or are severely damaged sometime later, by a traumatic event such as intrauterine stroke. Intrauterine strokes can lead to the re-absorption of the damaged cerebral tissue, and the replacement of this tissue by cerebrospinal fluid—in which case the anencephaly in question can be described as *hydranencephaly* (Merker 2007a: 78). To the extent we agree that hydranencephalic children can correctly be described as enjoying some level of phenomenal experience, we can safely conclude that the mind-body-self cannot be dependent on the cerebral cortex.

5.2.1.1 Shewmon et al.'s Cases

One of the most important studies on hydranencephaly is Shewmon et al.'s (1999) *Consciousness in congenitally decorticate children: developmental vegetative state as self-fulfilling prophecy*. These authors' main concern is that hydranencephalic children are routinely, but prematurely, classified in the diagnostic category 'persistent vegetative state'—probably because of our ingrained habit of thinking that the cerebral cortex is the seat of consciousness. The Medical Task Force on Anencephaly, as recently as 1990, stated that "[i]nfants with anencephaly, lacking functioning cerebral cortex, are permanently unconscious." Moreover, since "the suffering associated with noxious stimuli (pain) is a cerebral interpretation of the stimuli," it followed that "infants with anencephaly presumably cannot suffer" (in Shewmon et al. 1999: 364).

But Shewmon et al. most categorically do not agree with the pessimistic view of the MTFA. In Shewmon et al.'s view, categorizing decorticate children as 'vegetative' is tantamount to a self-fulfilling prophecy: as a result of being viewed as 'permanently unconscious,' these children are often not provided with the rich and stimulating environment they would need in order to reach their full potential; therefore, it is not surprising that they do not develop optimally, and may remain 'vegetative.'

The children in Shewmon et al.'s study, in contrast, displayed normal sleep/wake cycles, and would comfortably qualify as 'conscious' on the Glasgow Coma Scale (which is the standard instrument for determining the level of consciousness of coma patients). Shewmon et al. report that the four children in their review had "congenital brain malformations involving total or near-total absence of cerebral cortex," but still showed discriminative awareness: "for example, distinguishing familiar from unfamiliar people and environments." They were capable of "social interaction, functional vision, orienting, musical preferences, appropriate affective responses, and associative learning" (Shewmon et al. 1999: 364). It only takes a single counterexample to disprove a universal statement, and the authors think that any one of the four cases they present is a sufficient disproof of the claim that all hydranencephaly patients are 'permanently unconscious.'

Subject one's CT scan detected no cerebral tissue at all rostral to the thalamus, with the exception of a thin crescent of tissue in the mesial temporal lobe. "EEGs [electroencephalograms] showed no electrocerebral activity over the entire head" (ibid.) except for a tiny bit of low-amplitude wave activity in this area of remaining tissue. Although the cortical regions required for vision were atrophied, in both left and right hemispheres, this subject reacted to bright light at six weeks of age; by six years, he blinked when objects approached his eyes, closed his eyes in bright light, and was able to follow the approximate path of a moving object. At two years old, he could push himself along whilst lying on his back, and habitually moved through the glass sliding door (if it was open, which he could detect) onto the porch outside, to lie in the sun. From three years old, the patient reacted differentially to different toys, indicating not only that he could tell the difference between them, but that he had consistent preferences. He also reacted differentially to different types of music. If somebody called to him, he turned towards the person calling, and smiled.

Subject 2 was a classic case of hydranencephaly. Light was able to 'transilluminate' her skull, since there was no tissue whatsoever to block it (Shewmon et al. 1999: 366). All that was left was a very thin fibrous membrane, on both sides of the inferior aspect of the temporo-occipital region. From early in life, she could distinguish light from dark (becoming agitated when left in the dark, and unable to go to sleep until the lights were switched back on), and from about four years old, she could track objects visually at least some of the time. She could recognize her mother, and showed consistent preferences for certain types of music.

Subject 3 had "a gross brain malformation mainly resembling hydranencephaly but partially alobar holoprosencephaly" (the failure of the embryonic forebrain to divide into two hemispheres, Shewmon et al. 1999: 368). The only brain tissue she had spanned both left and right halves of the skull, without the normal division into hemispheres. Unfortunately, an overly pessimistic report from her initial paediatrician led to the parents' more or less giving up on this child; but when she started therapy with a new paediatrician, she immediately started to become less irritable and more responsive. This subject would reliably look up and smile when called, but could differentiate between strangers who were and were not comfortable around her. In one case, she cried throughout the visit of a certain therapist whom her mother described as 'condescending.' If someone was holding a toy, and suddenly whisked it out of sight, she clearly looked around to see where it had gone. She even learned the names of a few of the common household items around her, to the extent that if she was asked 'Where is [one of the objects]?' she looked towards the correct thing.

Like subject 2's, subject 4's skull 'transilluminated': the classic indication of hydranencephaly. He had severe optic atrophy, and showed "some response to dark and light but no visual tracking." The only part of his brain that showed any electrical activity at all was the occipital region (important for visual processing)—but this activity took the form of "frequent epileptic discharges" (Shewmon et al. 1999: 369), rather than the kind of activity associated with normal brain function. At about two years old, the patient could distinguish his mother; at nine years old, he was observed cooing, expressing sadness or pain, and smiling when his caregivers were around. By age 17, he used different facial expressions to indicate preferences.

5.2.1.1.1 Shewmon et al. on Cortical Remnants

So, as we can see, hydranencephalic children certainly seem to show some level of 'consciousness,' even the type of consciousness that is routinely measured on the Glasgow Coma Scale; they can even retain some not-insignificant perceptual-cognitive abilities. Not surprisingly, given the strong tendency in twentieth-century cognitive sciences to assume that the cortex was essential for consciousness and 'higher' (perceptual) functioning [Sect. 5.1.1.2], many have tried to account for the preserved abilities of hydranencephalic children by suggesting that some small remnants of cortex are actually still intact enough to be able to function. However, as Shewmon et al. are quite right to insist, "the point is that even if [subject 1 and subject 3] were not decorticate absolutely, they were enough so that physicians, including neurologists, predicted a vegetative outcome absolutely" (1999: 371). Obviously, it makes no sense to say both that a patient is no better than a vegetable, because of his/her lack of cortex; and that the patient has enough surviving remnants of cortex to enable rudimentary vision (that is, that the patient *is* significantly better than a vegetable, and *does not* completely lack cortex). Anybody who thinks that a hydranencephalic child is 'vegetative' is logically debarred from also suggesting that the child has enough surviving cortex to support any kind of perceptual or cognitive abilities.

Moreover, even if there was a little bit of intact cortical tissue remaining in some or all of these cases, those cortical structures the children did have could not plausibly account for their 'consciousness.' Subjects 1 and 3, who had surviving "rudimentary limbic structures" showed more affect, as well as more sociable behavior, than the other two—which is what one would expect, since the (subcortical) limbic system is known to be associated with emotion. But in spite of their total lack of visual cortex in the occipital region, these two subjects also had better vision than subjects 2 and 4. It would not be very plausible to suggest that the very minimal "mesial temporal tissue" and "frontal sliver" of cerebral cortex remaining in subjects 1 and 3 somehow got to have surprising idiosyncratic capabilities, and thus produced better vision in these subjects, even though vision is not something these areas are known for producing. Rather, the authors suggest, it would be "much more parsimonious … tentatively to ascribe subject 3's visual function … to subcortical pathways … rather than to her rudimentary frontal lobe" (1999: 372); and it is difficult to find fault with this reasoning.

Suggesting that the visual abilities of subject 1 and 3 were related to the intact subcortical brain structures of these subjects is more plausible than suggesting that they were related to the surviving remnants of cortical tissue they had. It is plausible to suggest that the hydranencephaly patients' remaining subcortical structures were sufficient for supporting both (easy-problem) 'conscious' perceptual-cognitive capacities and (hard-problem) 'conscious' phenomenal experience; therefore, we can deduce that the cerebral cortex is not essential for 'consciousness,' of either the easy-problem or hard-problem variety.

5.2.1.2 Merker's Cases

The other main influential hydranencephaly study is Bjorn Merker's (2007a) *Consciousness without a cerebral cortex: A challenge for neuroscience and medicine* (which is published together with a comprehensive list of commentaries, some of which I will briefly discuss below [Sect. 5.3]).

Merker's primary concern in his article is to introduce "a novel principle relating target selection, action selection, and motivation" (2007a: 63), and discuss how target selection, action selection and motivation could be carried out by subcortical mechanisms. This focus is significantly different from the main focus of Shewmon et al.'s article; however, on the question of the label 'persistent vegetative state' being a self-fulfilling prophecy, all the authors seem to have a similar view. Merker notes that "[i]t is not uncommon for parents to encounter surprise on the part of medical professionals when requesting analgesia or anesthesia for their crying child during invasive procedures, a situation in some ways reminiscent of what was found in the case of neonates only a few decades back" (2007a: 80).

Merker, like Shewmon et al., does not agree with this pessimistic view of hydranencephalic children. In the first months (up to about a year) of life, he writes, hydranencephalic children do not necessarily show any symptoms, and the condition may only be diagnosed when parents become worried about the children missing developmental milestones. When symptoms do start to appear, they "invariably" include motoric ones; quite often, this results in the children's being institutionalized (which can be seen as a self-fulfilling prophecy [compare: Sect. 5.2.1.1]). When the various health-related complications are controlled, however, the children "tend to stabilize and … can survive for years and even decades" (2007a: 79).

As long as such children are in a medically stable condition, and they are provided with the right kind of stimulation, "they give proof of being not only awake, but of the kind of responsiveness to their surroundings that qualifies as conscious by the criteria of ordinary neurological examination" (ibid.). They not only respond to environmental stimuli (especially auditory stimuli, but also visual ones), but respond *differentially* to different stimuli [Sect. 5.2.1.1]. They clearly experience (and display) basic feelings of pleasure and unpleasure, and their characteristic displays of pleasure and unpleasure can consistently be elicited by those who are familiar with the children.

The case of absence epilepsy in hydranencephalic children is another good piece of evidence that they should be considered 'conscious' (or, as I would prefer to put it, that these children enjoy rudimentary affective awareness). During these absence episodes (which are similar to the 'absences' displayed by Anna O., Sect. 4.2.2.1), the children are simply 'not *there*.' They do not respond to things they usually respond to, nor seem to be aware of anything going on around them. When they are 'back' again, their caregivers can clearly perceive the difference. Absence seizures are to some extent similar to slow-wave sleep, in that, while they are having these seizures, children are much less conscious of their surroundings and much less responsive to sensory stimulation. However, also like slow-wave sleep, absence seizures can be terminated by particularly powerful stimuli; moreover, like slow-wave sleep, absence seizures do not seem to preclude the possibility of "unconscious stimulus evaluation" (Coenen 2007: 88). 'Relevant' stimuli can rouse both deep sleepers and absence epilepsy sufferers more easily than non-relevant stimuli (for example, if somebody softly speaks the name of a sleeper—or of somebody who is in the throes of an absence episode—that person is much more likely to wake up than if somebody softly says any other word; compare also 'the cocktail party effect,' Sect. 3.3.2.1).

This suggests that, in both deep sleep and hydranencephaly, the subjects are 'conscious' in some minimal sense, at least enough to evaluate the relevance of some sensory stimuli. In Merker's opinion, the evidence from cases of hydranencephaly "gives no support for … the assumption that 'awareness,' in the primary sense of coherent relatedness of a motivated being to his or her surroundings, is an exclusively cortical function."

> The tacit consensus concerning the cerebral cortex as the 'organ of consciousness' would thus have been reached prematurely, and may in fact be seriously in error. (Merker 2007a: 80)

5.2.1.2.1 Merker on Cortical Remnants

Merker's cases, like Shewmon et al.'s cases, detract from the plausibility of suggesting that residual function in these children can be attributed to remnants of cortical tissue. Visual processing areas in the brain are often left relatively undamaged by hydranencephaly, whereas "it is rare for any auditory cortex to be spared" (Merker 2007a: 79)—and yet it is usually the opposite way round when it comes to sensory responsiveness. The children generally have more or less functional hearing, and it is vision which is more often and/or more severely compromised.

This paradoxical finding can be easily accounted for, however, by pointing to the undamaged auditory system in the brainstem the children would have. In contrast, the optic nerve—which is situated right near the cerebral cortex—would normally be damaged whenever the cortex itself was damaged (or, even more to the point, whenever the cortex was almost completely *absent*, as in hydranencephaly). That would explain why hydranencephalic children tend to have better hearing than vision.

Merker's cases, like Shewmon et al.'s cases, indicate that the cerebral cortex is not essential for easy-problem (perceptual-cognitive) or hard-problem (phenomenal/what-it-is-like) consciousness.

5.2.1.3 The Lesson (of 'Consciousness in Hydranencephaly') for Philosophy

Whereas Merker is interested in finding out empirically which specific brain structures are essential for which functions, I am more interested in the conceptual implications of the empirical findings. I am not concerned with whether or not children with certain kinds of deficits should be classified as 'vegetative,' or with whether or not it is plausible to suggest that certain cortical remnants could be responsible for the children's perceptual or cognitive abilities. All I want to do here is to show that, even though those children have little or no functioning cortex, they can still justifiably be described as 'conscious,' in both the easy-problem and the hard-problem sense. Accordingly: the corticocentric view is problematic, and we should not take the mind-body-self to be dependent on the functioning of the cerebral cortex.

Just as ('sensory') perceptual-cognitive (and/or introspective, and/or rational) 'consciousness' are only specific and rather limited instances of what-it-is-like/experience, so cortical consciousness is also a specific and rather limited type of it. In fact, to the extent that perception, cognition, introspection and rationality depend on the cerebral cortex, we might even say that perceptual-cognitive and/or introspective and/or 'rational' consciousness is *the same* rather limited type of consciousness as cortical consciousness.

5.2.2 Solms on Types of Consciousness, the Structure of the Brain, and the Structures of the Mind

In the previous chapter, I examined the first part of Solms' analysis of Freud's view: whether or not he had correctly identified Freud's 'two major mental systems' (including whether or not they corresponded to Solms' 'internal body' and 'external body'), and whether or not he had accurately reflected what Freud said about whether the id and the ego were 'conscious' [Sect. 4.3.2]. In both these respects, I determined that Solms' claims were not well-founded. But quite an important part of Solms' original challenge (in *The Conscious Id*, 2013) to Freud's view had to do with the way in which different types of 'consciousness' (and/or/therefore, the different structures of the mind) are produced in the brain.[3]

According to Solms,

> although there is in these [hydranencephalic] children significant degradation of the types of consciousness that are normally associated with representational perception and the cognition derived from it, there can be no doubt that they are conscious, both quantitatively and qualitatively. They are not only awake and alert, they also experience and express a full range of instinctual emotions. The primary (affective) self is, in short, *present*. (2013: 11)

Therefore, because these children have no cortex, and yet are conscious in the primary affective sense, Solms reasons that

> (i) affective consciousness is both generated and felt subcortically. (Solms 2013: 11)

Solms evidently takes it for granted that Freud believed 'consciousness' to be a cortical function, because he says that this "contradicts the theoretical assumptions of [various people, including] Freud" (ibid.). Moreover, he takes the belief that consciousness is a cortical function to be central to Freud's view. Therefore, the fact hydranencephalic children are (affectively) 'conscious,' even though they have no cortex, appears to him to present a strong challenge to the Freudian framework.

Solms also thinks his hydranencephaly cases can be used to show that, contrary to what Freud thought,

> (ii) consciousness is not inherently perceptual: it is inherently affective. (Solms 2013: 12)

Solms is apparently also taking it for granted that, on Freud's view, consciousness is inherently perceptual. In this case, the fact that the 'consciousness' displayed by the hydranencephalic children is of the affective variety would appear to count against Freud's view.

Finally, Solms assumes that these first two claims—that [i] 'affective consciousness is generated and felt subcortically' and that [ii] 'consciousness is not inherently perceptual; it is inherently affective'—imply that

> (iii) *consciousness is generated in the id*, and the ego is fundamentally unconscious. (Solms 2013: 12)

Clearly, Solms takes it for granted that, on Freud's view, the ego is *conscious*, and the id is *un*conscious (and in fact I have already explicitly ascribed such a reading to Solms, Sect. 4.3.2.2)—therefore, Solms assumes, if he is able to show that the ego is unconscious, and the id is conscious, that would severely undermine Freud's model of the mental apparatus.

However, I can see four different respects in which Solms' claims may be problematic.

5.2.2.1 The Structure of the Brain vs Types of Consciousness

The first thing we need to ask is: *Is* affective consciousness really generated and felt subcortically?

5.2.2.1.1 In Hydranencephaly

Solms, as we can see from the above passage (Solms 2013: 11), is willing to accept that

(a) Hydranencephalic children have (at best) only 'degraded' perceptual-cognitive consciousness; and
(b) Hydranencephalic children are nevertheless 'conscious' (both quantitatively and qualitatively).

Given these two points, Solms claims that the obvious conclusion to draw is the one he in fact draws:

(c) The consciousness enjoyed by hydranencephalic children must be of the (primary) affective variety (rather than the perceptual-cognitive variety).

Moreover,

(d) The affective consciousness experienced by these children must be generated and felt subcortically (there would be no possible way for it to be generated or felt in the cortex, since these children completely lack cortex).

That all seems to be satisfactory.[4]

5.2.2.1.2 In Freud

But even if we accept that affective consciousness is generated and felt subcortically, we also need to ask: *Is* the fact that affective consciousness is generated and felt subcortically in conflict with Freud's view?

According to Solms, Freud never questioned the classical assumption that 'consciousness' was a cortical function (Solms 2013: 9); and his evidence for this claim comes from a passage which I have already partially quoted in Chap. 4 [Sect. 4.3.2.2], namely:

> What consciousness yields consists essentially of perceptions of excitations coming from the external world and of feelings [which] arise from within the mental apparatus:

The passage continues:

> it is therefore possible to assign the system Pcpt.-Cs. ['perception-consciousness'] a position in space. It must lie on the borderline between outside and inside; it must be turned towards the external world and must envelop the other psychical systems. It will be seen that … we have merely adopted the views on localization held by cerebral anatomy, which locates the 'seat' of consciousness in the cerebral cortex—the outermost, enveloping layer of the central organ. (Freud 1920: 24, in Solms 2013: 9)

Solms claims that "it is clear from [this] quotation that even the internal aspect of consciousness was, for Freud, 'lodged on the surface of the brain'" (2013: 9)—but I cannot see that that is justified. What Freud said was that the system *Pcpt.-Cs.* was located in the outermost layer of the brain, which he took to be the cerebral cortex. He said that we could assign the system *perception-consciousness* a position in space; and then he said that, by locating the seat of *consciousness* in the cerebral cortex, he is doing no more than adopting the accepted views on cerebral anatomy. Although he does not specifically stipulate that it is *perceptual* consciousness which must be 'seated' in the cerebral cortex, it seems clear from the previous sentences that in all these cases Freud is considering 'consciousness' exclusively in the perceptual sense, and *not* 'the internal aspect of consciousness' referred to by Solms. As I explained [Sect. 5.2], virtually all contemporary experts agree that both perception and 'consciousness' mostly depend on the cerebral cortex: therefore, Freud's claim that perceptual consciousness is 'seated' in the cerebral cortex is unproblematic. Even in the early twentieth century, evidently, this was already the accepted view.

In which case: even if we acknowledge that affective (or 'internal') consciousness is generated and felt subcortically, this is not in conflict with Freud's claim that *perceptual* consciousness was a cortical function. I have already mentioned [Sect. 4.3.1.2] that, according to Freud, 'internal' perceptions (for example, feelings of pleasure/unpleasure) 'can come about even when 'consciousness' is clouded.' Therefore, it seems reasonable to suggest that Freud might describe the hydranencephaly patients as experiencing 'internal perceptions,' such as pleasure/unpleasure—as being 'conscious' in this very limited sense—while their perceptual (and/or introspective and/or cortical) 'consciousness' was 'clouded.'

In this case, the claim that 'Consciousness is both generated and felt subcortically' does not appear to be in conflict with anything Freud said. This means that Solms' claim [Sect. 5.2.2], that the case of hydranencephaly threatens Freud's view (with regard to the structure of the brain vs types of consciousness), is problematic.

5.2.2.2 The Essential Nature of Consciousness

Next, we need to ask: *Is* consciousness inherently affective?

5.2.2.2.1 In Hydranencephaly

Hydranencephalic patients in some cases lack all perceptual and cognitive capacities, but as Solms says, there is 'no doubt' that they are 'conscious.' The sense in which they are 'conscious' seems to be similar to the sense in which, I suggested, Freud would have allowed that the id was 'conscious' [Sect. 4.3.2.2]: the patients are 'conscious' in the rudimentary affective sense, but *not* in the perceptual-cognitive and/or introspective sense. As I said above [Sect. 4.3.2.2.2], it makes no sense to talk about 'feeling' something 'unconsciously.' Seeing that hydranencephaly patients may show this rudimentary affective type of awareness, even when they lack all perceptual capacities, and/or capacities for introspection, it seems plausible to claim that the rudimentary affective type of awareness is more fundamental than perceptual-cognitive and/or introspective/rational awareness. There are some cases (such as hydranencephaly) in which only rudimentary affective awareness is present; there are some cases in which rudimentary affective awareness and more advanced perceptual-cognitive and/or introspective-rational awareness are both present (in 'normal' people); but there do not seem to be cases in which perceptual-cognitive and/or introspective-rational awareness is present while rudimentary/affective awareness is absent. The rudimentary affective type of awareness seems to be prerequisite for the other type.

In which case: saying 'Consciousness is inherently affective' is unproblematic.

5.2.2.2.2 In Freud

I established that it was plausible to read Freud's claim that the id was deeply/permanently unconscious as a claim about perceptual (and/or introspective) consciousness—which therefore did not rule out the possibility of saying that the id was 'conscious' in the rudimentary/affective sense of being aware of its own sensations of pleasure and unpleasure [Sect. 4.3.2.2]. It seems that, if the 'consciousness' *of the id* is of this affective variety, then—seeing that the id is the main constituent of the mental apparatus (and that the other structures of the mind are just 'specially modified' parts of the id, Sect. 4.3.1.3)—it would be reasonable to say that 'consciousness' *tout court* must be inherently affective. Seeing that the id is the main constituent of the mental apparatus, and the ego (and superego) is only a specially modified part of the id, then it is problematic to claim that the ego (or superego) is 'conscious,' in any sense, unless the id is also conscious, in at least some sense.

In other words, 'Consciousness is not inherently perceptual, it is inherently affective' does not appear to be in conflict with Freud's view. This means that Solms' claim, that the case of hydranencephaly threatens Freud's view (with regard to the essential nature of consciousness), is problematic.

5.2.2.3 Types of Consciousness vs the Structures of the Mind

Whereas Solms thinks that (ii) 'consciousness is not inherently perceptual, it is inherently affective,' and (iii) 'consciousness is generated in the id, and the ego is fundamentally unconscious' [Sect. 5.2.2] are two separate claims, I have already suggested that it is precisely *because* Freud considered the consciousness of the id to be of the affective variety that it was reasonable to take his view to be that consciousness *tout court* was affective [Sect. 5.2.2.2]. On my reading, saying that 'consciousness is generated in the id' is virtually identical to saying that 'consciousness is inherently affective'—on the understanding that (a) the type of consciousness intrinsic to the id is the affective type [compare: Sect. 4.3.2.2], and that (b) the id comprises the bulk of the mental apparatus (the ego and superego being 'specially modified' parts of the id) [Sect. 4.3.1.2]. In this case, seeing that I have already explained why the claim (ii) that 'consciousness is not inherently perceptual, it is inherently affective' does not constitute a challenge to Freud's view, then, to about the same extent, the claim that (iii) 'consciousness is generated in the id, and the ego is fundamentally unconscious' also does not constitute a challenge to it.

5.2.2.4 The Intended Scope of 'Cortical'

As I said [Sect. 5.2.2.1], Solms determined that, on Freud's view, consciousness was a cortical function—on the strength of a remark made by Freud about locating the seat of consciousness in the cerebral cortex, in accordance with the prevailing views on cerebral anatomy. According to Solms, Freud states his view "even more explicitly" (2013: 9) in *An Outline of Psycho-Analysis*. Freud remarks that even the 'internal' perceptions of the interior of the body "emanate from the terminal organs" of the body; and Solms assumes that, because Freud said that "we regard all these as prolongations or offshoots of the cortical layer" (Freud 1940: 161–162, in Solms 2013: 9), the implication is that on Freud's view, even 'internal' perceptions/feelings must be produced cortically. But this is problematic. To be sure, Freud repeatedly claimed, in many different works, that Pcpt-Cs was a 'cortical' function—but I argue that it is a mistake to take these remarks to commit Freud very strongly to any views about literal, physical brain structures.

In *The Unconscious*, Freud remarked that neurological research was starting to indicate that specific brain regions might be linked to the functioning of specific parts of the body, and particular mental activities; but that every attempt to draw firm conclusions about exactly *which* brain regions were linked to *which* mental activities "has miscarried completely" (1915, in *Complete Works*, p. 2998). This suggests that, when he referred to the 'cortical' layer of the brain, Freud may not have intended to make any very firm claim about a literal brain structure (as opposed, for example, to intending to make a more functional claim). This seems particularly plausible in light of the fact that he describes the cerebral cortex as 'the outermost, enveloping layer of the central organ' [Sect. 5.2.2.1]: it is whatever brain structure is situated at the intersection of 'inside' and 'outside'.[5]

In his *Introductory Lectures on Psycho-Analysis*, Freud made some remarks which also seem to be highly pertinent here:

> We are told [for example] that the medulla oblongata is stimulated, and the patient learns that he is suffering from a neurosis of the vagus nerve. The medulla oblongata is a very serious and lovely object. I remember quite clearly how much time and trouble I devoted to its study many years ago. To-day, however, I must remark that I know *nothing that could be of less interest to me for the psychological understanding of* [for example] *anxiety than knowledge of the path of the nerves along which its excitations pass.* (1917, in *Complete Works*, p. 3442, my emphasis)

When Freud said that 'the seat of consciousness is located in the cerebral cortex' [Sect. 5.2.2.1], my guess is that (just as in the case of anxiety) *absolutely nothing* hangs on whether or not 'consciousness' really has anything to do with the literal cerebral 'cortex' (as in, the outer layer of the cerebral hemispheres). We should take 'cortical' here as a functional descriptor, and not an anatomical descriptor. In which case, it is problematic for Solms to criticize Freud's view regarding where in the brain 'consciousness' was produced, insofar as this implies that Freud even *had* any firm view about brain structures and their functions.

Finally: Solms identifies the following passage, in which (he says) Freud even more explicitly states his 'corticocentric' view:

> The process of something becoming conscious is above all linked with the perceptions which our sense organs receive from the external world. From the topographical point of view, therefore, it is a phenomenon which takes place in the outermost cortex of the ego. (Freud 1940: 161–162, in Solms 2013: 9)

But it looks to me as if Solms has missed the key significance of this passage with regard to Freud's views on the cortex: according to Freud, the process of something becoming conscious takes place 'in the outermost cortex *of the ego*,' and not 'in the outermost cortex *of the brain*'! Since the ego is not a physical structure (which is extended in space), I do not believe that it would make any sense to read 'outermost cortex [of the ego]' as implying any kind of anatomical commitment on Freud's part regarding the actual cerebral cortex of the physical brain. And since, as we know, the ego (and superego) are really specially modified parts of the id, we can conclude that if the conceptual 'location' of the ego in the cerebral cortex is more functional than anatomical, then the 'location' of the other structures of the mind should likewise be taken in this conceptual sense, rather than a literal, anatomical sense.

Therefore, Solms' claim, that the case of hydranencephaly threatens Freud's view (with regard to Freud's supposed views about 'cortical' functioning), is problematic.

5.2.2.5 The Lesson (of 'Solms on Consciousness and the Brain') for Philosophy

Clearly, there is no reason to disagree with Solms' claim that affective consciousness is generated and felt subcortically. However, since Freud only ever said that *perceptual* consciousness was a cortical phenomenon, the fact that affective consciousness is generated and felt subcortically does no damage to Freud's framework [Sect. 5.2.2.1]. Nor is there any reason to doubt that 'consciousness' is inherently affective, rather than inherently perceptual, as Solms says. However, if I was correct to suggest that Freud differentiated between perceptual consciousness, in the cortex/ego, and rudimentary affective consciousness, in the id [Sect. 4.3.2.2], then this also does not constitute a problem for his framework. If we can correctly say that the id is conscious in the rudimentary affective sense, this is tantamount to saying that 'consciousness' *tout court* is inherently affective, since the id is the largest and most important part of the mental apparatus [Sects. 5.2.2.2 and 5.2.2.3].

Moreover, we cannot accept Solms' assertion that any of his observations about hydranencephaly are in conflict with any of Freud's views about the functioning of the brain—insofar as we might not be justified in thinking that Freud even *had* any views (in a strictly literal/anatomical respect) about the functioning of the brain [Sect. 5.2.2.4].

In fact, if I have correctly interpreted Freud's view of the mental apparatus, it seems probable that, had he had a better understanding of cerebral anatomy than he actually did, he himself would probably have said something very much along the same lines as what Solms is saying: the brainstem (and/or/therefore id) is 'unconscious' in the perceptual-cognitive/introspective sense, but *conscious* in the rudimentary affective sense of being 'conscious' of its own feelings. The ego (and/or/therefore cortex) is predominantly conscious, but contains some important unconscious contents. The hydranencephalic children, who are in possession of functioning brainstems, but not cortex, can be considered to have affectively-conscious, subcortical ids, but (in at least some cases) no perceptually-/introspectively-conscious, cortical egos.

Therefore (Solms' concerns notwithstanding), Freud's views on the different types of consciousness, the structure of the brain, and the structures of the mind seem to provide an excellent theoretical framework for understanding not only the respect in which 'what is psychical' can be dissociated from 'what is conscious' [Sect. 4.2], but also the respect in which intrapsychic conflict results from the simultaneous presence within the psychical apparatus of (un-/conscious) logically incompatible motives [Sect. 4.3]. On top of all that, Freud's account even performs well in light of recent neurological findings [Sect. 5.2.2].

5.2.3 The Cerebral Cortex and Consciousness: Conclusion

The discussion of hydranencephaly supports a number of claims I have made up to this point:

First, it supports both my claim that we should be careful not to conflate *consciousness* with ('conscious') perception or cognition [Sect. 3.2], and my claim that we should be careful not to conflate *what-it-is-like/experience* with (perceptual-cognitive) 'consciousness' [Sect. 3.3]. The hydranencephalic children qualify as 'conscious' on a normal neuropsychological diagnostic measure, and they indisputably enjoy a (rudimentary) kind of subjective, psychological what-it-is-like/experience, in spite of the fact they have very limited capacities for both perception and cognition.

Second, it supports both my claim that we should be careful not to conflate (psychological) what-it-is-like/experience with (introspective) consciousness [Sect. 4.2], and my claim that we should be careful not to conflate (psychological) what-it-is-like/experience with rationality [Sect. 4.3]. These children indisputably enjoy a (rudimentary) kind of psychological what-it-is-like/experience, in spite of the fact they have very limited capacities for both (introspective) consciousness and rationality.

Finally, this discussion of hydranencephaly supports my first new claim in Chap. 5 [Sect. 5.2] that we should not take the mind-body-self to be dependent on the functioning of the cerebral cortex—because the type of rudimentary affective awareness enjoyed by the children (who have little or no functioning cerebral cortex) seems to provide the foundation for all other types of awareness or experience.

I can now complete my assessment of Solms' case against Freud. With regard to Freud's two main psychical systems, and to whether the id/ego are conscious/unconscious, I have already determined that nothing Solms said constituted either a challenge to, or an improvement on, what Freud said [Sect. 4.3.2]. In addition, with regard to the types of consciousness which can most properly be ascribed to the functioning of the subcortical and/or cortical brain; with regard to the essential nature of consciousness; with regard to the types of consciousness attributable to the different structures of the mind; and with regard to the sense in which Freud (may have) used the term 'cortical,' I have now determined that Solms has not succeeded in improving on Freud's own views [Sect. 5.2.2]. There is nothing about the brain or its operation which prevents us from taking a Freudian-type view of a mind-body-self which is neither essentially (introspectively) conscious, nor essentially rational—nor/therefore, essentially cortical.

Because hydranencephalic children enjoy rudimentary affective awareness—in spite of the fact that they have little or no functioning cerebral cortex—we should not take the mind-body-self to be dependent on the functioning of the cerebral cortex. Since Freud only ever said that *perceptual* consciousness was dependent on cortical functioning, this does not conflict with anything in the psychodynamic framework for the mental apparatus (in spite of Solms' claims to this effect).

5.3 Rudimentary Affective Awareness

For the purposes of my argument in Chap. 3, I defined 'affective' as 'relating to, arising from, or influencing feelings or emotions' [Sect. 3.2.3.2.1]. In this section, I need a slightly refined definition. Panksepp understands the kind of rudimentary affective awareness which he calls 'primary-process consciousness' as "*that ineffable feeling of experiencing oneself as an active agent in the perceived events of the world*" (1998: 310), or the "subjective experiential-feeling component [of awareness] ... that is very hard to describe verbally" (2005a: 32). From now on, I will take this definition as epitomizing the type of rudimentary affective awareness that I am interested in: that ineffable feeling of experiencing oneself in the world which is very difficult to describe.

The problem with exclusively studying a specialized form of 'consciousness,' such as adult, human, 'hard-problem' consciousness, is that what appear to be 'universal' features could turn out to be specific to that particular specialized form of consciousness under study. Merker suggests that, to avoid this risk, the best course would be consider as many different kinds of 'consciousness' as possible; "to first abstract a common denominator of conscious states from their many forms (2007a), and let what they share rather than what divides them inform our conception of the nature of conscious function" (Merker 2007b: 117, see also Panksepp 1998, 2012; Watt 2005; Liotti and Panksepp 2004: 65).

And 'the common denominator of conscious states' appears to be a kind of rudimentary affective awareness, which provides "an essential psychic scaffolding" (Panksepp 1998: 309) for any other, 'higher' forms of 'consciousness'.[6] However much we may think we humans have risen above the level of mere brutes, "our internal feelings will always be guided by … our deep affective reactions to events in the world" (Panksepp 1998: 320). Therefore, the 'common denominator' of different kinds of 'conscious states' should be taken to be basic affective states; and investigating basic affective states should be taken to be an essential first step towards finding out anything about 'higher' human, introspective 'consciousness.'

That seems reasonable; and in fact is much the same premise I relied on when I was discussing Solms' analysis of Freud's views [Sect. 5.2.2]. I said that, if the consciousness *of the id* was subcortical and/or affective, then it would be reasonable to assume that consciousness *tout court* was subcortical/affective—since the ego and superego were only 'specially modified' parts of the id. Similarly, if the common denominator of 'consciousness' is *basic affective states*, as both Panksepp and Merker are suggesting, then it would be reasonable to assume that understanding the essential nature of basic affective states might provide good insight into 'consciousness' *in general*—since our uniquely human, reflective 'consciousness' is really just a specially modified type of rudimentary affective awareness.[7]

Now that I have determined that we should be careful neither to conflate 'consciousness' [Sect. 3.2] with ('conscious') perception/cognition, nor to conflate *what-it-is-like/experience* [Sect. 3.3] with (perceptual/cognitive) *consciousness*; that we should be careful not to conflate (psychological) what-it-is-like/experience with (introspective) consciousness or rationality [Sects. 4.2 and 4.3]; and that we should understand that psychological what-it-is-like/experience is (therefore) not a question of cortical functioning [Sect. 5.2]—I am going to consider some of the commentaries to Merker's article [Sect. 5.3.1]. These commentaries support my claim that investigating the kind of rudimentary affective awareness which is present in hydranencephaly is a good first step towards understanding the (human) mind-body-self: that we *should* understand the mind-body-self as essentially affective.

5.3.1 Merker Commentaries

In order to understand why living creatures have the capacities they do, it is necessary to understand how those capacities evolved from the simpler capacities of simpler creatures which preceded them in evolution. One of the main tenets of the 'affect-centric' view (which I will be discussing in depth in the next chapter, but which is similar to the views expressed by Merker and Panksepp [Sect. 4.3]) is that

> [i]f basic forms of consciousness emerged in ancestral species that predated humans …, an understanding of the neural instantiation of those processes … is bound to inform us profoundly about the nature of our own conscious abilities. (Panksepp 1998: 34)

And first thinking about basic forms of consciousness in ancestral species is exactly what Merker does. Right at the beginning of his article, he briefly discusses the cubomedusa, a type of predatory jellyfish:

> The four semi-independent pacemakers of the non-encephalized nervous system of the cubomedusa equip this predatory jellyfish with flexible directional locomotor responsiveness to asymmetric sensory inputs … There is no reason to assume that the environmental guidance thus supplied by its

> radially arranged nerve net involves or gives rise to experience of any kind. Our own environmental orientation, on the other hand, commonly takes place in a state of wakefulness we call conscious, which typically involves seeing, hearing, feeling, or other kinds of experience. (Merker 2007a: 63)

Merker claims that cubomedusas, with their simple nervous systems, very likely do not enjoy any kind of 'experience'—and, if this is the case, it seems that there is no conceptual reason that even much more complex creatures (like humans, for example) should not be able to handle the day-to-day requirements of staying alive quite satisfactorily without 'experience' either. The fact that at least some creatures *do* appear to enjoy a kind of ('conscious') experience "suggests that consciousness has some role or function to fill in the neural economy of brains thus endowed" (2007a: 63) (bearing in mind, of course, that we need to insert some mental scare quotes around 'consciousness').

It might be reasonable to say that the hydranencephalic children in the studies I mentioned have rudimentary affective awareness, despite the fact that they (or at least some of them) have no perceptual-cognitive capacities, or capacities for (introspective) 'consciousness' or rationality. Depending on exactly how 'conscious' is understood, it might even be reasonable to say that they have rudimentary affective awareness despite the fact that they *are not* 'conscious' at all. Freud said that it was profitless to dispute whether preconscious ideas could more properly be said to belong to the system *Conscious.* or the system *Unconscious.* [Sect. 4.2.1.3]; similarly, we might decide that whether or not the rudimentary type of affective awareness enjoyed by hydranencephaly patients should qualify as a kind of 'consciousness' boils down to nothing more than a profitless verbal dispute.

In philosophy, we would probably refer to the rather limited kind of consciousness displayed by hydranencephalic children as 'creature consciousness'; namely, "whatever differentiates ordinary people who are either awake or in REM sleep from ordinary people who are in non-REM sleep, in a coma, and so forth" (Piccinini 2007: 103). According to Piccinini, most philosophers think of creature consciousness as a kind of necessary, but wholly uninteresting, prerequisite for the more interesting types of perceptual/cognitive (and/or introspective/rational) consciousness that they like philosophizing about. In contrast, Merker's article gives us more than enough reason to think that "part of what makes a system have experiences [in the first place] is that it is creature-conscious" (ibid.). The commentaries on Merker's article comprise a range of views on 'primary process consciousness,' 'core consciousness,' and 'affective consciousness' (all of which are similar to 'creature consciousness')—and this type of consciousness is of the utmost interest and importance to these researchers.

5.3.1.1 The Ontology of Creature Consciousness: A Challenge for Philosophy (Gualtiero Piccinini)

Piccinini agrees with Merker that the best way of finding out about human phenomenal what-it-is-like/experience is to start with more basic forms of awareness that creatures (including both cubomedusas and hydranencephalic children) might have in common. As I mentioned, Merker's main concern in his article was to discuss a novel principle of target selection, action selection, and motivation; he discussed the role of 'primary consciousness' in 'integrating sensory information and motivations to select targets and actions' [Sect. 5.2.1.2], but he conspicuously did not talk about "what it takes *for a mental state* to be conscious" (Piccinini 2007: 103). He did not even use the term 'mental state' anywhere in his article. So we could anticipate that some philosophers would immediately dismiss the article, claiming that, in view of his failure to discuss mental states, Merker has not even shown that the decorticate children are 'conscious' in the first place, let alone provided an acceptable account of 'consciousness.'

Alternatively, though, Piccinini suggests, we could try relinquishing some of our preconceptions, and take some form of primary affective awareness as the starting point for some new investigations. Whereas most philosophers do not see 'creature consciousness' as very interesting or important [compare: Sect. 5.3.1], Merker's article, according to Piccinini, gives us more than enough reason to think that this may be an unenlightened way of looking at things. He thinks that attempting to account for Merker's decorticate children in terms of 'state consciousness' is just the wrong way of tackling the issue—because "part of what makes a system have experiences [in the first place] is that it is creature-conscious" (ibid.).

> If creature consciousness were at least partially constitutive of phenomenal consciousness, it would be a mistake to develop theories couched solely in terms of state consciousness, without saying anything about creature consciousness—as philosophers are fond of doing. (2007: 104)

Therefore, a satisfactory account of human (phenomenal/hard-problem) consciousness should begin with creature consciousness. Insofar as Merker is not concerned with state consciousness, and insofar as he seems to take creature consciousness seriously, Merker's account is satisfactory to Piccinini.

I completely agree with that; moreover, I said much the same thing myself [Sect. 5.2.3]: I said that the task before us was to investigate the subjective experiential-feeling component of (rudimentary affective) awareness in more depth. Insofar as the kind of creature consciousness discussed by Piccinini is similar to what I have been calling 'rudimentary affective awareness,' Piccinini's claim that a correct account of consciousness should begin with creature consciousness amounts to much the same thing as my claim that we *should* understand the (subcortical) mind-body-self as essentially affective.

5.3.1.2 Levels of Emotion and Levels of Consciousness (Carroll Izard)

Izard thinks that the claims Merker makes about the children in his study "present a strong challenge to prevailing assumptions about the primacy of cognition in emotion-cognition-behavior relations," and (therefore) also constitute "a strong case for the upper brain stem as being the neural home of primary or phenomenal consciousness" (2007a: 96).

While immediate, pre-reflective awareness of feelings may not be sufficient to qualify as 'consciousness' (depending on exactly how 'consciousness' is understood), it *is* enough to generate "the basic emotion of interest" (2007a: 97): a rudimentary positive or negative attitude towards something which is or is not valuable to the survival of the organism. In both humans and other animals, the basic emotions of interest include excitement, joy, and anger. Moreover (according to Izard), having interests is just as important for very primitive kinds of organisms (like cubomedusas) as it is for the most sophisticated. Even single-celled organisms have the capacity to detect and approach stimuli that are in some sense valuable to them (for example, moving out of environments with the wrong temperature, towards environments with the right temperature; moving towards nutrients); and being motivated in the two basic directions of approach or withdrawal (that is to say, having 'interests') is also the basis of even much more complex emotions.

Having interests is clearly not one of the characteristics that depends on the highly evolved (human) neocortex, and 'the basic emotions of interest' could therefore be precisely the kind of emotions which are displayed by the hydranencephalic children.[8] These basic emotions are therefore not the kind of thing that should be understood in terms of perception/cognition, (introspective) consciousness/rationality, and/or/therefore the cerebral cortex.

On Izard's view, "basic emotions [such as 'interest'] constitute the motivational system ('bias') in the processes of primary consciousness." Whereas reflective consciousness is characterized by "symbolic processes, memory, and, ultimately, the capacity for awareness of self and others," primary consciousness is characterized by "sensory processes that generate subjective feelings … especially emotion feelings" (ibid.). The advantage of saying that emotional feelings are the central component of 'primary' consciousness is that this makes it possible to say that a creature is 'conscious' of a feeling, without being able to label or articulate that feeling, or having the requisite self-awareness to be able to think of itself as having that feeling.

Therefore, a satisfactory account of human (phenomenal/hard-problem) consciousness should take basic emotions into consideration. Insofar as Merker seems to take basic emotions seriously, his account is satisfactory to Izard.

> Attributing emotion feeling to primary consciousness means that one can become conscious of a feeling that one cannot label and articulate. (2007a: 97)

Insofar as Izard thinks that basic emotions are the key ingredients of primary consciousness, his view supports my claim that we *should* understand the (subcortical) mind-body-self as essentially affective.

5.3.1.3 Affirmative-action for the Brainstem in the Neuroscience of Consciousness: The Zeitgeist of the Brainstem as a 'Dumb Arousal' System (Douglas Watt)

As I said [Sect. 5.1.1.2], there has been not only a widespread assumption that consciousness must be a cortical function, but also a corresponding widespread assumption that the brainstem and other sub-cortical structures are only useful for the functions of basic 'biological regulation.' This view of brainstem function is "a begrudging acknowledgement" (Watt 2007: 108) of Moruzzi and Magoun's seminal work which showed that brainstem damage led to somnolence and coma: the view acknowledges that the brainstem is not only essential for producing and maintaining biological life, but is also a prerequisite for 'consciousness.' However, proponents of this view think that it is not the subcortical structures of the brain themselves which carry out the 'higher' functions (like 'consciousness' et cetera): they merely *enable the cortex* to carry out those higher functions.

On such a view, subcortical structures can be seen as constituting a kind of "dumb arousal system," whose primary (if not *sole*) responsibility is 'waking up' the cortical brain structures: "the brainstem does for the forebrain [that is, the cerebral cortex and some adjacent structures] essentially what a battery does for a light" (Watt 2007: 109). Coenen (also in his commentary on Merker) offers an excellent example of a 'dumb arousal' view. He says that two systems working together are responsible for producing 'consciousness': "the midbrain [that is, 'subcortical'] system for waking and vigilance (the engine), and the ... cortical system for perception and experience (the vehicle)" (Coenen 2007: 88). The midbrain structures are needed to give the cortical system what it requires, *in order to* produce consciousness—but it is the cortical system which actually *produces* the consciousness. Hypothetically speaking, on such a view, any other system that helped the cortical system to maintain its wakeful and vigilant state could quite satisfactorily replace the midbrain/subcortical structures which are in practice responsible for this function.[9]

However, according to Watt, it is absurd to think that such a simplistic framework as the dumb arousal view could ever do justice to "the functional complexity of the brainstem" (2007: 109). In his opinion, trying to claim that all the vital, complex, and interconnected functions of the subcortical brain can be reduced to something as simplistic as 'arousal' may not even be completely *coherent*. The dumb arousal view says that the brainstem is necessary for putting the cortex into the necessary state of arousal, and that it is the cortex which then produces consciousness. But if 'a state of arousal' refers to 'a state of the organism in which stimuli elicit appropriate functional responses,' then saying that the brainstem is responsible for producing such a state is tantamount to saying that *the brainstem itself* is what produces those appropriate, functional responses. If certain brainstem functions are required to put the cortex into the requisite state of arousal it needs to be in before the cortex can produce appropriate functional responses, it is difficult to see why those requisite brainstem functions should be considered as unimportant or uninteresting, at the same time that the cortical responses which rely on those brainstem functions are considered both interesting and important.

There is possibly a sense in which the brainstem functions have to precede the cortical functions temporally—but does the fact that the cortical functions always come (maybe 'a few split-seconds') *after* the brainstem functions necessarily indicate that they are more important, or in some other way 'better'? On the contrary, could the fact that the brainstem functions always *precede* the cortical functions not be taken as evidence that the brainstem functions are prerequisite for (and therefore 'more important than') the cortical functions?

Even the most basic creatures (like for example the cubomedusa) need sensory systems for mapping images of potential food, motor systems for "mapping trajectories to the food," and "a homeostatic representation of metabolic shortfall" (ibid.) to tell them when the shortfall has been successfully remedied. But it would not serve any organism very well to be in possession of these systems if the systems were not "in register with one another." And, as Merker has so convincingly argued, it is the mesencephalon (in the subcortical brain) which synthesizes the input from these systems. Therefore, Watt thinks it might be more accurate to characterize the function of the mesencephalon in terms of the integration of motor, sensory, and homeostatic functions—and to think of brainstem mechanisms as "smart integration systems" (ibid.), rather than 'dumb arousal' systems.

> If the extended group of reticular systems [in the brainstem] enables coherent and purposeful behaviors to emerge from the system, then it cannot simply be 'turning on the lights' but must be underpinning a global integration of function of the kind Merker outlines. (2007: 109)

Therefore, a satisfactory account of human (phenomenal/hard-problem) consciousness should not propose that the brainstem has any kind of 'dumb arousal' function. Insofar as Merker seems to recognize the complexity and importance of the function of the subcortical brain, his account is satisfactory to Watt.

I said that both Shewmon et al. and Merker seemed to think that hydranencephalic children are prematurely judged to lack the capacities for 'consciousness' or for 'suffering' [Sect. 5.2.1]. No doubt Watt would agree on this point, since on his view the capacities of the brainstem (which are the only capacities which children with no cortex can possess) are complex and sophisticated. Insofar as Watt thinks that the brainstem is responsible for 'global integration of function,' as well as rudimentary affective awareness, his view supports my claim that we should not take the mind-body-self to be dependent on the functioning of the cerebral cortex [Sect. 5.2], and my claim that we *should* understand the (subcortical) mind-body-self as essentially affective [Sect. 5.3].

5.3.1.4 Raw Feeling: A Model for Affective Consciousness (Van Honk et al.)

As I explained in Chap. 3, mind sciences researchers trying to investigate 'consciousness' often focus on 'the easy problems' of consciousness (namely, 'conscious' perception and cognition) at the expense of 'the hard problem' [Sect. 3.1]. They attempt to "reduce consciousness to measurable properties"; for example, "studying neural correlates of sensory awareness, such as meaningless randomly moving dots" (Van Honk et al. 2007: 107). And, while there is a certain sense in which it is interesting to find out what brain processes are necessary before we can become 'conscious' of sensory perception, at the very best, this is an extremely limited kind of 'consciousness.' Living creatures do not live in laboratories populated by meaningless computer-generated images: they live in the real world—and in the real world, "'raw feelings' mediate conscious adaptive responses" (ibid.).

Seeing that the characteristics any living creature has must (by definition) have been advantageous to that organism, in terms of survival fitness, in order to have endured, there must be a good explanation for how rudimentary affective awareness confers evolutionary advantage.[10] Van Honk et al. suggest that if the organism can experience "raw feelings of reward and punishment," it will be able to "make fine-tuned decisions of approach or withdrawal-related action" (ibid.), which will help it survive. The basic decision-making tool that any successful creature (even a cubomedusa) needs to have is what the authors call a "detection-evaluation-decision (DED) device," to enable it to determine what is going on in the environment, how what is going on in the environment affects it, and what to do about it. But this DED differs greatly across the different phylogenetic levels.

In the most primitive ('instinctive reptilian') type of brain, "DED processing occurs at an instinctual brainstem level"; in the more complex 'emotional paleomammalian' brain, "the DED system copes with threat by initiating fight/flight behaviors" in the amygdala/hypothalamus (a little closer to the cortex than the brainstem is, but still 'subcortical'); and finally, in the highly sophisticated 'cognitive neomammalian' brain, "higher-order [cortical] cognitive faculties such as reasoning and language" (ibid.) are added to the DED process.

If the 'detection-evaluation-decision' function is the main respect in which 'consciousness' confers evolutionary advantage, then this same (or at least, a very similar) function must be present in much more rudimentary types of 'consciousness' than adult humans typically enjoy, and which preceded adult human consciousness in evolution. Therefore, studying the form of ancestral consciousness which preceded our own—the rudimentary affective awareness of the emotional paleomammalian brain—can be expected to be very valuable with regard to achieving a fuller understanding of the human mind-body-self.

Accordingly: a satisfactory account of human consciousness should take important primitive forms of consciousness, such as 'raw feelings,' into consideration. Insofar as Merker seems to take raw feelings seriously, his account is satisfactory to van Honk et al.

> Consciousness evolved to ensure adaptive homeostasis … The mechanism relies on the subject's capacity to experience raw feelings of reward and punishment. (2007: 107)

On van Honk et al.'s view, Merker's study confirms that the affective consciousness of the subcortical brain provides the foundation for the more sophisticated consciousness of the cortex [compare: Sect. 5.3]. Insofar as van Honk et al. think that it is important to study earlier forms of consciousness in order better to understand our own, this commentary supports my claim that we *should* understand the (subcortical) mind-body-self as essentially affective.

5.3.1.5 Emotional Feelings Originate Below the Neocortex (Jaak Panksepp)

Recall what I said about Churchland: she assumed that the task for 'neurophilosophy' was (among other things) to find out 'what makes animals smart' [Sect. 3.1], and to come up with 'new theories about the nature of information-processing' [Sects. 2.3.2.1 and 3.1]. In contrast, Panksepp denies both that 'consciousness' has anything to do with being smart, and that it is a question of any kind of information-processing (2007: 102). Rather, he understands 'consciousness' as "the experience of body and world, without necessarily understanding what one is experiencing." Moreover, Panksepp thinks that the best way to investigate 'secondary' and 'tertiary' types of consciousness (that is, "the ability to have thoughts about the world" and the ability to have "thoughts about thoughts") is first to accomplish a good understanding of the more basic, more affective (and subcortical) kind of 'consciousness' shared even by much simpler creatures than ourselves [Sect. 5.3].

Primary phenomenal states, for Panksepp, comprise "two distinct but highly interactive branches: (1) the ability to perceive and orient in the world, and (2) the ability to feel the biological values of existence." Of these 'branches,' Panksepp thinks Merker has focused more on the former, while he himself is more interested in the latter. Core motivational and emotional processes, in Panksepp's view, had a more important role to play than perceptual processes during "the emergence of experiential capacities within brains" (ibid.) in evolution. Core motivational and emotional processes correspond to bodily values: essentially, these basic positive and negative emotions motivate organisms—even very primitive organisms, such as cubomedusas—to seek out or to avoid whatever it is that elicits them.

Panksepp takes it to be uncontroversial that when "practically all higher-brain regions are removed in animals … or congenitally absent in human children [for example, in hydranencephaly], core consciousness survives" (ibid.). Primary consciousness is the product of "somato-visceral operating systems of the upper brainstem" (Watt and Pincus 2004, in Panksepp 2007: 102)—Panksepp describes it as a "deeply personal" type of awareness, relating to an organism's bodily needs. He believes that "the more ancient … motivational-emotional urges of the brainstem were foundational for the more lateral zones that harvest external information for the guidance of behavior" (2007: 102)—and as I said [Sects. 5.1.1.2 and 5.2], there is a growing tide of empirical evidence supporting this claim. As the corticocentric fallacy loses its hold, more and more people are starting to think that brainstem structures are critical not only for 'dumb arousal,' but for plenty of other 'vital, complex, and interconnected functions'—including, presumably, such 'motivational-emotional urges' as Panksepp refers to.

Merker's data (not to mention, Shewmon et al.'s data, Sect. 5.2.1.1) supports Panksepp's view about the primacy of rudimentary affective awareness over perceptual-cognitive awareness (and the primacy of the subcortical brain over the cortex), insofar as "[p]rimary consciousness in [the] neurologically impaired children was most dramatically evident in their affective presence" (ibid.). The hydranencephalic children, in spite of their severe deficits, had more or less intact motivational and emotional capacities (as well as some rudimentary perceptual capacities)—on account of their functioning brainstems.

> If we wish to scientifically understand the nature of primary-process consciousness, we must study the subcortical terrain where incredibly robust emotional and perceptual homologies exist in all mammalian species. (2007a: 103)

Therefore, a satisfactory account of human (phenomenal/hard-problem) consciousness should begin with an account of primary/affective consciousness; and a correct account of primary/affective consciousness should begin with an account of the subcortical brain. Insofar as Merker seems to take seriously the centrality both of rudimentary affective awareness and of the subcortical brain, his account is satisfactory to Panksepp.

Panksepp's view provides particularly strong support for my suggestion that we should be careful not to conflate (hard-problem) what-it-is-like/experience with (easy-problem) perceptual or cognitive 'consciousness,' (introspective) consciousness or rationality, and/or/therefore the functioning of the cerebral cortex. Instead, we should focus on the type of rudimentary affective awareness exhibited by these children. We *should* understand the (subcortical) mind-body-self as essentially affective.

5.3.1.6 The Lesson (of the Merker Commentaries) for Philosophy

Although almost all the commentators think that Merker's article is important and valuable, several of them also criticize Merker for his "neglect of the obvious and massive contribution of the telencephalon [that is, the left and right hemispheres of the cerebral cortex, and some subcortical structures directly adjacent to them] to adult human consciousness" (Merker 2007a: 110). However, this criticism is undeserved. Merker explicitly stated that the point of the article was not to find out how adult human consciousness was instantiated by brain structures: he was interested in investigating which types of creatures (including human creatures with certain types of deficits) could be considered 'conscious,' and in determining whether investigations of very primitive types of 'consciousness' might have interesting or important implications for "consciousness more generally" (Merker 2007a: 111).

Of the commentators, Merker says that Piccinini seems to have understood his main point more or less correctly; the one thing he is not quite happy with is Piccinini's sharp differentiation between the type of 'consciousness' experienced by children with and without hydranencephaly. Merker thinks that the cases *are* different, but that the difference does not have to do with 'state consciousness' as opposed to 'creature consciousness' (which is what Piccinini assumes). Merker thinks that the philosophical distinction between state consciousness and creature consciousness is just that: a *philosophical* distinction. From an empirical point of view, he sees little justification for drawing this distinction.

What Merker and most of the commentators do seem to agree on is that investigating (subcortical) rudimentary affective awareness could be a valuable way of finding out more about 'the hard problem' of consciousness (that is, subjective/psychological what-it-is-like/experience)—even the *much* more advanced type of (introspective) 'consciousness' encountered in normal adult humans.

5.3.2 Rudimentary Affective Awareness: Conclusion

Ralf-Peter Behrendt (2007) suggests that it is not plausible to say that children who show the rather limited kind of awareness to the environment that I've been describing should be called 'conscious'—especially when "purposefully reacting invertebrates, such as the medusa … are not." But it seems to me that Behrendt is the one who is being implausible here, and this is because the children in the study not only behave appropriately in the presence of certain stimuli, but also appear to experience these stimuli (especially in an affective sense): it's not just that they are able to move around objects they can't 'see' without bumping into them, but also that they are able to search for hidden toys, display idiosyncratic musical tastes, recognize familiar people, and be afraid of the dark. I would say that this is precisely the point that Merker is making.

I think it is fair to say that both Merker's article, and these commentaries on his article, support my claim that we should not take the mind-body-self to be dependent on the functioning of the cerebral cortex [Sect. 5.2]—because it is not plausible to deny that the hydranencephalic children described by Shewmon et al. and by Merker enjoy some kind of ('conscious') phenomenal experience. Moreover, both Merker's article and these commentaries on his article support my second claim (that we *should* understand the mind-body-self as essentially affective [Sect. 5.3])—because the type of rudimentary affective awareness enjoyed by the hydranencephalic children, who have little or no functioning cerebral cortex, seems to provide the foundation for all other types of awareness or experience. Some of those children have virtually no perceptual and/or cognitive capacity, and are neither (introspectively) conscious nor rational—and yet the one thing they appear to experience very intensely is affective feelings. In fact, Freud's view [Sects. 4.2 and 4.3], Solms' view [Sects. 4.3.2 and 5.2.2], Merker's/Shewmon et al.'s views [Sect. 5.2.1], *and* the views of the commentators I have discussed here [Sect. 5.3.1] all support my claim that we should understand the essential nature of the mind-body-self in terms of non-perceptual-cognitive, non-(introspectively)conscious, non-rational, rudimentary *affective awareness*.

Because rudimentary affective awareness can be present when 'consciousness' is absent, we should understand perceptual-cognitive/introspective-rational consciousness as derivative, and rudimentary affective awareness as the most fundamental form of 'consciousness.' We *should* take the (subcortical) mind-body-self—which is not essentially perceptual, cognitive, (introspectively) conscious, or rational—to be essentially affective.

5.4 The Brain and the Mind-body-self: Conclusion

So at this point, I have demonstrated (in Chap. 2) that we should not talk about the explanatory gap between 'the mind' and 'the body,' since we are not dualists. I have demonstrated (in Chap. 3) that—even if we temporarily overlook the fact that the reification of 'consciousness' is just as bad as the reification of 'mind'—we should be careful not to conflate either 'consciousness' [Sect. 3.2] or what-it-is-like/experience [Sect. 3.3] with ('conscious') sensory perception or cognition. Not only is quite a lot of perception and cognition actually *not* conscious, but non-conscious perception/cognition is also (therefore) not associated with any what-it-is-like/experience. I have demonstrated (in Chap. 4) that we should also be careful not to conflate (psychological) what-it-is-like/experience with (introspective) 'consciousness' [Sect. 4.2], since the existence and significance of (dynamically) *un*conscious psychological material can be both theoretically accounted for and empirically corroborated; moreover, since some of the most influential factors contributing to both psychological and somatic experience are logically incompatible with one another, we know that we should be careful not to conflate psychological (what-it-is-like) experience with 'rationality' [Sect. 4.3].

In Chap. 5, I have now also demonstrated that:

(i) Because hydranencephalic children enjoy rudimentary affective awareness—in spite of the fact that they have little or no functioning cerebral cortex—we should not take the mind-body-self to be dependent on the functioning of the cerebral cortex [Sect. 5.2] (which is not in conflict with anything Freud said).

My discussion of hydranencephaly has also supported my first positive claim in this book:

(ii) Because rudimentary affective awareness can be present when 'consciousness' is absent, we *should* take the (subcortical) mind-body-self—which is not essentially perceptual, cognitive, (introspectively) conscious, or rational—to be essentially affective [Sect. 5.3].

Accordingly, the next thing I am going to do is to examine in depth two specific examples of views which take affect seriously.

Notes

1. In other words, Freud, *at this very early point in his career*, held a dualistic view.
2. Ironically, focusing a disproportionate amount on the wise and subtle deliberations of the cortex is exactly what I'm going to accuse Damasio himself of doing [Sect. 6.2.1].
3. Although I maintain that no true physicalist can recognize the existence of such 'thing' as 'consciousness,' I need to use the term in a scare-quotes sense in this section, as I did in Chaps. 2 and 3, in order to evaluate Solms' argument.
4. According to Gallese (2013: 43), Solms' view is problematic insofar as it "downplays too much the role played by the neocortex in a variety of aspects of conscious life." However, as I argue throughout Sect. 5.2, there is good reason to think that the role of the neocortex in 'conscious life' has been exaggerated.
5. Compare: Sect. 4.2.1.2—the 'surface' of the 'undifferentiated vesicle' would come to embody whichever qualities allowed it best to perform its primary function of interfacing between the interior and the exterior of the vesicle.
6. I will discuss some of this research further in the next chapter.
7. Although, technically speaking, Merker is still committing reification, when he says 'a common denominator of *conscious states*'—as is Panksepp, when he says 'higher forms of consciousness'—I am inclined to forgive them this slight lapse. Nothing could be clearer than that they are *not* thinking of 'consciousness' in the sense of a 'thing' which has causal

power. That subjective (psychological) what-it-is-like/experience I am interested in is clearly exactly what Merker and Panksepp are also interested in.

8. For Izard (2007b: 272), the basic emotion of interest is intrinsically conscious, just as feelings/affects are intrinsically conscious for Freud [Sect. 4.3.2.2].
9. I mentioned [Sect. 5.1] that distinguishing between 'the cerebral cortex' and 'the subcortical brain' was a gross oversimplification which I was intentionally in the name of clarity; although Watt discusses some specific subcortical/brainstem structures, and their differential contributions to 'consciousness,' that level of detail is similarly not required for my purpose in this chapter.
10. Strictly speaking, the trait must *originally* have conferred some advantage (although it may no longer do so), or must in some (other) way have been *associated* with some trait that conferred some advantage.

References

Baars, B.J. 2005a. Global workspace theory of consciousness: Toward a cognitive neuroscience of human experience. *Progress in Brain Research* 150: 45–53.

———. 2005b. Subjective experience is probably not limited to humans: The evidence from neurobiology and behavior. *Consciousness and Cognition* 14: 7–21. https://doi.org/10.1016/j.concog.2004.11.002.

Behrendt, R.-P. 2007. The hypthalamo-tectoperiaqueductal system: Unconscious underpinnings of conscious behavior. *Behavioral and Brain Sciences* 30 (1): 85–86. https://doi.org/10.1017/S0140525X07000933.

Cacioppo, J.T., and G.G. Berntson. 2004. Social neuroscience. In *The cognitive neurosciences*, ed. M. Gazzaniga, 3rd ed., 977–985. ProQuest Ebook Central database: https://0-ebookcentral-proquest-com.ujlink.uj.ac.za/lib/ujlink-ebooks/detail.action?docID=3338826.

Coenen, A.M.L. 2007. Consciousness without a cortex, but what kind of consciousness is this? *Behavioral and Brain Sciences* 30 (1): 87–88. https://doi.org/10.1017/S0140525X07000957.

Damasio, A. 1994. *Descartes' error: Emotion, reason and the human brain*. New York: Avon Books.

Freud, S. 1895. *Project for a scientific psychology*. Unpublished manuscript. http://users.clas.ufl.edu/burt/freud%20fleiss%20letters/200711781-013.pdf. Accessed 20 September 2015.

———. 1915. The unconscious. In *Complete Works*, pp. 2989–3024.
———. 1916–1917. Introductory lectures on psycho-analysis. In *Complete Works*, p. 3124.
———. 1953. *On Aphasia*. New York: International Universities Press.
Freud—Complete Works, Ivan Smith 2000, 2007, 2010. https://www.valas.fr/IMG/pdf/Freud_Complete_Works.pdf. Accessed 20 September 2015.
Gallese, V. 2013. Bodily Self, Affect, Consciousness, and the Cortex. *Neuropsychoanalysis* 15 (1): 42–45. Available from Vittorio Gallese (https://www.academia.edu/4401896/Gallese_V._2013_Bodily_self_affect_consciousness_and_the_cortex._Neuropsychoanalysis_15_1_42-45?auto=download).
Gazzaniga, M.S., R.B. Ivry, and G.R. Mangun, eds. 2002. *Cognitive neuroscience: The biology of the mind*. 2nd ed. New York: W. W. Norton.
Izard, C.E. 2007a. Levels of emotion and levels of consciousness. *Behavioral and Brain Sciences* 30 (1): 96–98. https://doi.org/10.1017/S0140525X07001045.
———. 2007b. Basic emotions, natural kinds, emotion schemas, and a new paradigm. *Perspectives on Psychological Science* 2 (3): 260–280.
Kandel, E.R. 2005. *Psychiatry, psychoanalysis and the new biology of mind*. Washington: American Psychiatric Publishing.
Liotti, M. & Panksepp, J. 2004. Imaging human emotions and affective feelings: implications for biological psychiatry. In *Textbook of biological psychiatry*: 33–74. Edited by Panksepp, J. Available from ProQuest Ebook Central database: https://0-ebookcentralproquest-com.ujlink.uj.ac.za/lib/ujlink-ebooks/reader.action?docID=469784.
Merker, B. 2007a. Consciousness without a cerebral cortex: A challenge for neuroscience. *Behavioral and Brain Sciences* 30: 63–81. https://doi.org/10.1017/S0140525X07000891.
———. 2007b. Grounding consciousness: The mesodiencephalon as thalamocortical base. *Behavioral and Brain Sciences* 30: 110–134. https://doi.org/10.1017/S0140525X07001173.
Moruzzi, G. & Magoun, H.W. 1949. Brain stem reticular formation and activation of the EEG. *Electroencephelography and Clinical Neurophysiology* 1: 455–473.
Panksepp, J. 1998. *Affective neuroscience: The foundations of human and animal emotions*. ProQuest Ebook Central database: https://0-ebookcentral-proquest-com.ujlink.uj.ac.za/lib/ujlink-ebooks/detail.action?docID=431176#.
———. 2001. The neuro-evolutionary cusp between emotions and cognitions: Implications for understanding consciousness and the emergence of a unified mind science. *Evolution and Cognition* 7 (2): 141–163.

———. 2003. At the interface of the affective, behavioral, and cognitive neurosciences: Decoding the emotional feelings of the brain. *Brain and Cognition* 52: 4–14. https://doi.org/10.1016/S0278-2626(03)00003-4.

———. 2005a. Affective consciousness: Core emotional feelings in animals and humans. *Consciousness and Cognition* 14: 30–80. https://doi.org/10.1016/j.concog.2004.10.004.

———. 2007. Emotional feelings originate below the neocortex: Toward a neurobiology of the soul. *Behavioral and Brain Sciences* 30 (1): 101–103. https://doi.org/10.1017/S0140525X07001094.

———. 2012. A synopsis of affective neuroscience—Naturalizing the mammalian mind. *Journal of Consciousness Studies* 19 (3–4): 6–18.

Piccinini, G. 2007. The ontology of creature consciousness: A challenge for philosophy. *Behavioral and Brain Sciences* 30 (1): 103–104. https://doi.org/10.1017/S0140525X07001100.

Redding, P. 2000. Freud's theory of consciousness. In *The analytic Freud: Philosophy and psychoanalysis*, ed. Michael Levine, 101–118. London: Routledge.

Shewmon, D.A., G.L. Holmes, and P.A. Byrne. 1999. Consciousness in congenitally decorticate children: Developmental vegetative state as self-fulfilling prophecy. *Developmental Medicine & Child Neurology* 41: 364–374. https://doi.org/10.1111/j.1469-8749.1999.tb00621.x.

Smith, D.L. 1999. Sigmund Freud's programme for a science of consciousness. *British Journal of Psychotherapy* 15 (4): 414–424.

Solms, M. 2013. The conscious id. *Neuropsychoanalysis: An Interdisciplinary Journal for Psychoanalysis and the Neurosciences* 15 (1): 5–19. https://doi.org/10.1080/15294145.2013.10773711.

Solms, M., and J. Panksepp. 2012. The "id" knows more than the "ego" admits: Neuropsychoanalytic and primal consciousness perspectives on the interface between affective and cognitive neuroscience. *Brain Sciences* 2: 147–175.

Van Honk, J., B.E. Morgan, and D.J.L.G. Schutter. 2007. Raw feeling: A model for affective consciousness. *Behavioral and Brain Sciences* 30 (1): 107–108. https://doi.org/10.1017/S0140525X07001148.

Watt, D.F. 2005. Panksepp's common sense view of affective neuroscience is not the commonsense view in large areas of neuroscience. *Consciousness and Cognition* 14: 81–88. https://doi.org/10.1016/j.concog.2005.01.003.

———. 2007. Affirmative-action for the brainstem in the neuroscience of consciousness: The zeitgeist of the brainstem as 'dumb arousal' system. *Behavioral and Brain Sciences* 30 (1): 108–109. https://doi.org/10.1017/S0140525X07001161.

6

Twenty-first-century Biological Naturalism: The Body-Map-Based View and the Affect-Centric View

6.1 Introduction: The Fall of Behaviorism

On the behaviorist view, the behaving organism was seen as a 'black box' whose internal workings were both mysterious and unimportant. So long as there was *something* that reliably produced certain behavioral output in response to environmental input, it made little difference, for the behaviorists, whether or not it was possible to say *how* or *why* that response was produced in that way [Chap. 5: The brain and the mind-body-self]. But "[t]hat extreme level of skepticism was so unrealistic, and so at variance with everyday experience, that it was abandoned in most branches of psychology" (Panksepp 1998a: 10). Just as disenchantment with psychodynamics helped pave the way for behaviorism to become dominant [Sect. 4.1], so, eventually, disenchantment with behaviorism set the scene for 'the cognitive revolution.'

During the cognitive revolution (in about the 1950s and 60s), computational theories of mind started to exert more and more influence on psychology. Although it might not seem that an increased focus on cognition or computation would be likely to generate an increased interest in 'the hard problem' of consciousness, to some extent, this is exactly what happened. The new cognitive trend was responsible for lifting the taboo

© The Author(s), under exclusive license to Springer Nature Switzerland AG 2022

J. Anderson, *Biological Naturalism and the Mind-Body Problem*,
https://doi.org/10.1007/978-3-030-99684-0_6

on mental processes which had been imposed by behaviorism; and after the cognitive revolution, mental processes once again became the subject of respectable discourse. However, just as the behaviorists were completely uninterested in finding out anything about brain function, so (at least the first few waves of) the cognitive scientists tended not to devote much attention to principles of brain functioning. On any computational view, as on any behaviorist view, it makes little difference what it is in the brain that carries out the relevant 'computations,' or exactly how that thing carries out those computations—so to that extent, (early) cognitive science was not much of an improvement over behaviorism.

In the Freudian era, psychiatrists focused on the deep dynamics of the mind—not so much *ignoring* the functioning of the brain as just *being unable to get to grips with* it, because of the fledgling state of neuroscience knowledge. Now, after the cognitive revolution, we have almost exactly the opposite problem: "[m]any neuroscientifically inclined descendants of those [behaviorist] hard-liners continue to advance the view that everything organisms do can be explained simply by an ultra-reductionistic neural analysis" (Panksepp 2005: 41)—without recourse to any psychological concepts. We focus on the functioning of the brain, and devote little (if any) attention to how brain functioning relates to (psychological) what-it-is-like/experience, and "throughout the twentieth century and until quite recently, both neuroscience and cognitive science gave emotion a very cold shoulder" (Damasio 1999: 383). But in the twenty-first century, "the cognitive revolution is gradually giving way to an emotion revolution" (Panksepp 2003b: 5).

This final chapter comprises a detailed comparison of the views of two of the great pillars of this emotion revolution, Antonio Damasio and Jaak Panksepp. Damasio [Sect. 6.2] is a renowned neuroscientist and neurobiologist, whose interests include emotions, feelings, decision-making and consciousness; Panksepp [Sect. 6.3] is a renowned neuroscientist and neuropsychologist, whose research has largely focused on analyzing the neuroanatomical/neurochemical mechanisms of emotional behaviors and feelings which are common to humans and other mammals. Although neither of these authors appears to have much interest in the problem of the explanatory gap, specifically, their views both qualify as 'biological naturalism': they embrace the claim that everything there is physical (or material, compare: Sect. 2.2.2)—and therefore that 'the mind' (or what-it-is-like/experience) is physical. Both Damasio and Panksepp offer accounts of how thinking/consciousness (or what-it-is-like/experience) is non-mysteriously produced during the course of the functioning of physical structures.[1]

According to Panksepp (1998a: 117), we have now reached the stage where we are able to account for psychological processes in terms of 'specifiable material particles,' just as Freud attempted to do [Sect. 4.2.1.1]. Maybe we could even say that Damasio's and Panksepp's 'biological naturalist' views are two specific examples of how to account for (psychological) what-it-is-like/experience in terms of specifiable material particles. However, in my opinion, Damasio's view places an inappropriate amount of emphasis on perception, cognition, introspection, rationality, and therefore, the functioning of the cerebral cortex. As I have explained in Chaps. 3, 4, and 5, this is problematic for anyone who is interested in what-it-is-like/experience. In contrast, Panksepp's 'affect-centric' view explicitly rejects many of the commitments which are central to Damasio's view; so, to the extent that the commitments of Damasio's view are problematic, and to the extent that Panksepp's view rejects these commitments, Panksepp's view should be preferred by anyone who is interested in what-it-is-like/experience.[2]

Panksepp takes it for granted that Freud's view of a self comprising a rudimentary affective kind of awareness, and emphatically not (introspective/reflective) 'consciousness' or rationality, is substantially correct, which is another factor which seems to me to count strongly in favour of Panksepp's view.

6.1.1 Useful Definitions

Damasio's view and Panksepp's view have both important similarities and important differences. Adding to the difficulty of comparing their two views is the fact that they sometimes use different terms for the same thing, and/or the same term for different things. So the first thing I need to do is clarify exactly what they take some key terms to signify.

6.1.1.1 Consciousness, Mind and Self

Damasio tells us that even very simple organisms can be counted as 'awake' (or maybe 'sentient'): having a brain is clearly not essential for sentience. What separates ('sentient') creatures with brains only from creatures with both brains and minds is that creatures with minds have "the ability to *display* images internally and to *order* those images in a process called thought" (1994: 89) (although these 'images' can be produced by any sensory modality, not just vision). Moreover, saying that a creature has 'a mind' "is the same as saying that that creature has cognition or cognitive processes" (1994: 90).[3]

'Minds' can be either 'conscious' or 'nonconscious.' *Consciousness*, for Damasio, requires an "organization of mind contents centered on the organism that produces and motivates those contents." That is the whole point of his calling the book *Self Comes to Mind*:

> [C]onscious minds arise when a self process is added onto a basic mind process. When selves do not occur within minds, those minds are not conscious in the proper sense. (2010: chap. 1, sect. 2)

The crucial point about this is that Damasio is effectively equating 'consciousness' with '*reflective self*-consciousness.' His view, therefore, is very restrictive: it severely limits the types of what-it-is-like/experience that can be can be referred to as 'consciousness.' Most people would probably be comfortable saying that there is something-it-is-like to be one of those hydranencephalic children [Sects. 5.2.1 and 5.2.2]—maybe even to be a cubomedusa [Sect. 5.3.1]—whereas we would be much less willing to allow that they were 'conscious' in Damasio's sense of having a self process added onto a basic mind process.

Panksepp believes that phenomenal consciousness can exist, and be studied, in the absence of any reflective recognition on the part of the organism that it is conscious. On his view, there is no important ontological difference between being able to display and order images internally and (also) having a sense of self as a protagonist among those images: for Panksepp, the rudimentary affective awareness displayed by the hydranencephalic children is just as valid a form of 'consciousness' as 'consciousness' of the reflective variety. Panksepp's view can therefore be described as very inclusive.

> Panksepp ... expands our notion of mind beyond the representational and propositional versions that dominate both cognitive science and traditional philosophy. (Asma and Greif 2012: 29)

6.1.1.1.1 Levels of Consciousness

Damasio distinguishes 'core' consciousness from 'extended' consciousness. 'Core' consciousness is "a simple, biological phenomenon [with] one single level of organization" (1999: 16). It is consciousness of the present moment (for example, 2010: chap. 7, sect. 5) which is not exclusive to humans (1998: 1880–1)—although it still requires some kind of a self process, added onto the basic mind process, and so is not available to very simple organisms. 'Extended' consciousness, on the other hand, is "a complex biological phenomenon [which] has several levels of organization; and ... evolves across the lifetime of the organism," incorporating the remembered past and the anticipated future. It is found in certain nonhuman species, but according to Damasio "it only attains its highest reaches in humans" (1999: 16)—that is, strictly speaking, it is not 'consciousness' that is uniquely human, but only 'the highest reaches of *extended* consciousness.'

Panksepp distinguishes between primary-, secondary-, and tertiary-process forms of consciousness. Primary-process consciousness reflects "the raw sensory/perceptual feelings and ... emotional/motivational experiences" (2005: 32) that even some of the simpler creatures (as well as hydranencephalic children, as we saw) might experience. Secondary consciousness is "the capacity to have thoughts about experiences, especially about how external events relate to internal events" (ibid.). Although we are reasonably confident that we are the only animals that think about our lives in linguistic terms, plenty of other animals might be capable of this kind of secondary consciousness in the form of sensory images. But tertiary-process consciousness—thoughts about other thoughts (2007a: 102)—is unique to humans (1998a: 35).

But crucially, for Panksepp, the fact that we humans are able to be aware of our own awareness in a 'tertiary' way is not *definitive* of anything. We already saw that, on Panksepp's view, all consciousness is based on 'primary-process consciousness,' which he understood as 'that ineffable feeling of experiencing oneself as an active agent in the perceived events of the world' [Sect. 4.3]. Whereas Damasio thinks that the highest reaches of extended (human) consciousness are very 'special' and 'unique,' Panksepp thinks that "secondary and tertiary forms of consciousness may be built integrally on more primary process variants" (2005: 63).

6.1.1.2 Emotions, Feelings, and Affects

Damasio uses *emotions* to designate "complex, largely automated programs of *actions* concocted by evolution"—or even "actions accompanied by ideas and certain modes of thinking" (2010: chap. 5, sect. 2). Although there are also cognitive aspects to these ('emotional') programs of actions, "the world of emotions is largely one of actions carried out in our bodies" (ibid.).

Feelings (of emotion) are "perceptions of what our bodies do during the emoting, along with perceptions of our state of mind during that same period" (ibid.). Whereas emotions are "outwardly directed and public," 'feelings' are "inwardly directed and private" (1999: 36). On Damasio's view, most 'feelings' take the form of variants of pleasure and 'unpleasure' [compare: Sect. 4.2.1.2]; but the word can also be used to refer to the sensation of touch, as in the feeling of the smoothness of velvet (2003: 3), as well as somatic feelings, such as the feeling of thirst.

In certain circumstances, especially those circumstances "when there is little or no time to examine feelings"—or, presumably, circumstances in which the creature having the feeling does not have sufficient cognitive sophistication to examine its own feelings—"feelings are solely the perception of a certain body state." In most other circumstances (at least, most circumstances involving human beings), there is also "a certain accompanying *mind state*" (2003: 89), which contributes to the feeling.

Damasio thinks that neural maps of the organism's internal state "form the basis for an important aspect of *the mental states known as feelings*"

(Damasio et al. 2000: 1051). Just as there were a number of types of what-it-is-like/experience which, on Damasio's narrow definition, would not count as 'consciousness,' so his understanding of 'feeling' is extremely narrow. We already know that he uses 'mind' synonymously with 'cognition'; now if he wants to say that *feelings* are a type of *mental state*, presumably the implication is that, on his view, organisms which do not have 'minds' are not only not capable of being 'conscious,' but also incapable of having *feelings*! In fact, he does explicitly make this claim: "the occurrence of a feeling in the traditional sense of the term require that its contents be known to the organism, that is, *consciousness is a requirement*" (2003: 110, my emphasis).

While it is reasonably uncontroversial that things like bacteria/cubomedusas et cetera might display 'emotional' behaviors (in the sense of largely automated programs of actions) in the absence of any corresponding internal feeling states, Damasio's insistence that a sense of self is necessary for being 'conscious' of feelings seems also to disqualify plenty of mammals—not to mention hydranencephalic children—that we might otherwise have been quite happy to consider as experiencing feelings.

Panksepp uses 'emotion' as a broad umbrella term which covers a wide range of "*affective*, cognitive, behavioral, expressive, and ... physiological changes" (2005: 32). He understands 'affect' as 'the subjective experiential-feeling component (of the various different changes that occur during emoting) that is very hard to describe verbally' [Sect. 5.3]; therefore, on Panksepp's view, "it is *a category error* to call affect ... unconscious" (2005: 61–2, my emphasis). Just as, I claimed [Sect. 3.3], there was *nothing*-it-is-like to 'perceive' something 'unconsciously' (or 'unconsciously' to carry out some cognitive operation), so there would be, for Panksepp, nothing-it-is-like to experience an affect 'unconsciously.' On Panksepp's view, "*all strong affects are intrinsically conscious*" (Watt 2017: 184)—just as, Freud claimed, 'an affective or emotional impulse is *necessarily felt*' [Sect. 4.3.2.2].

Another way of putting this point is to say that, on the affect-centric view, there is no difference between 'an emotion' and 'a feeling of emotion,' as there is on the body-map-based view.

6.1.1.2.1 Levels of Emotions

Damasio recognizes three categories of emotions: primary and secondary emotions (see below), and background emotions (which I will not discuss specifically). 'Primary emotions' are those emotions we first experience at a young age, and which therefore have limited cognitive components. Damasio suggests that it could easily be the case that "we are wired to respond with an emotion, in preorganized fashion, when certain features of stimuli in the world or in our bodies are perceived" (1994: 131). For example, it would be sensible if small creatures automatically and unthinkingly experienced fear when faced with large and threatening creatures—this would enable them to take immediate evasive action. It would not be necessary for the creatures to experience the emotion as a *feeling* of fear, as long as the emotion led them to exit the danger zone immediately.

But as we grow older, we experience primary emotions less often. Secondary emotions (or 'social emotions') include feelings "such as embarrassment, jealousy, guilt, or pride" (1999: 51), and we experience them "once we begin experiencing feelings and forming *systematic connections between categories of objects and situations, on the one hand, and primary emotions, on the other*" (1994: 134). Experiencing a secondary emotion comprises two steps. The first step involves deliberately thinking about the thing which triggered that emotion; such thinking likely comprises a procession of 'mental images,' both verbal and nonverbal. Seeing that this step requires the ability to display images internally (and manipulate the images during thought), this would not be possible for any creature not equipped with a *mind*.

The second step involves the brain's automatic, involuntarily reactions to the processing of the images. These involuntary reactions proceed in accordance with dispositions, which are acquired (again, automatically and involuntarily) during the course of many experiences of analogous situations. Seeing that the ability to acquire dispositions requires at least a minimal understanding of the relation of cause and effect [compare: Sect. 4.3.1], this second step would not be possible for any creature which did not have the capacity for a minimal level of rationality. As I explained in Chap. 4, all types of learning, rationality, cognition et cetera rely on the functioning of the cerebral cortex—which means that any kind of (secondary) 'emotion' that is understood as involving anything like learning, rationality or cognition must by definition also involve the functioning of the cerebral cortex.

Rather than distinguishing between primary and secondary emotions, as Damasio does, modern cognitive science (according to Panksepp) commonly distinguishes between 'first-order phenomenology' and 'second-order awareness.' This makes it possible, for example, to account for the Freudian concept of Repression [Sect. 4.2.2.2]: "even though people can often repress awareness of their emotional feelings at the level of second-order awareness, they may retain affective feelings at the level of first-order phenomenology" (Panksepp 2005: 40). Panksepp recognizes the validity of the distinction between 'first-order' and 'second-order' types of awareness, but from his point of view, the really important difference between the cases is not that one type of awareness is 'first-order' and the other type is 'second-order': it is that one is *affective*, while the other is *cognitive*. Saying that a certain type of awareness is 'affective' *necessarily implies* that that awareness is 'first-order.'

Whereas Damasio acknowledges that *primary* emotions may be unlearned, instinctive, and sub-cortical, Panksepp thinks that *all* 'emotions' are unlearned, pre-cognitive, and sub-cortical. That is how he *defines* 'emotions.'

6.2 Antonio Damasio: A 'Body-Map-Based' View of Human Consciousness and Human Reason

Damasio is one of the few neuroscientists in the twenty-first century to take emotion seriously. His popular books are not only important works in his own field, but also accessible enough for the non-specialist.[4] I can't possibly hope to do justice to all of Damasio's ideas, but I want to pick out some specific commitments of his view for the purposes of contrasting them with those of Panksepp's ideas which I discuss in 6.3.

6.2.1 The Central Topics

In the preceding chapters, I have discussed at length several ways of approaching 'the mind-body problem' which I consider to be unsatisfactory. I have argued that we should be careful not to conflate what-it-is-like/experience with ('conscious') perception or cognition; with (introspective) consciousness or rationality; and that we should not take the mind-body-self to be dependent on the functioning of the cerebral cortex. But it seems to me that Damasio does every one of these things I have claimed we should not do. On Damasio's view, the mind-body-self is conceptualized in terms of:

6.2.1.1 Perception and Cognition

6.2.1.1.1 Perception

As I explained, Damasio takes 'having a mind' as a question of 'being able to display images internally, and manipulate those images in the process of thought' [Sect. 6.1.1.1]. Everything we ever think about, by definition, takes the form of images: "[i]f [our thoughts] did not become images … they would not be anything we could know" (1994: 106). But although (as I said, Sect. 6.1.1.1) Damasio uses the word 'images' very broadly, to refer to the products of all sensory modalities, it seems clear that all kinds of 'images' must be things we *perceive*.

On Damasio's view, even *feelings* can be described as 'perceptual' in nature: they comprise primarily the perceptions of states of the body (as well as the accompanying mind states [Sect. 6.1.1.2]). Admittedly, the type of perceiving one does when perceiving one's own body state might be a somewhat different form of perceiving from the form involved in perceiving sense-data; but it is still the *perception* of this body state that concerns Damasio. Both having a mind, and having feelings, for him, are questions of different kinds of perceiving things through the senses.

6.2.1.1.2 Cognition

Damasio even goes so far as to say that "feelings are just as *cognitive* as other percepts" (1994: xv, my emphasis)! Feelings are a type of perception; perceptions are a type of cognition; the implication, therefore, is that all feelings, as well as all (other) perceptions, can all be described as 'cognitive.' I have just admitted that the type of perception involved in perceiving feelings might be different from the type involved in perceiving sensory stimuli; similarly, I would be happy to admit that the type of cognition involved in perceiving things was different from the type of cognition involved in paradigmatically cognitive activities. But even if perceiving feelings is not quite the same as other perceptual activities, and even if perceiving sense data is not quite the same as other cognitive activities, Damasio makes it very clear that, for him, feelings, perceptions, and cognitions can at least sometimes be very much the same sort of thing. Therefore, I feel justified in claiming that the body-map-based view is characterized by its emphasis on perception and cognition.

This is problematic from the point of view of anybody who is interested in what-it-is-like/experience, because, as I explained [Sect. 3.2], perception and cognition can often take place without generating any kind of what-it-is-like/experience. Moreover, I explained that individuals with little or no capacity for perception or cognition usually still enjoy a kind of rudimentary affective what-it-is-like/experience [Sect. 5.2].

6.2.1.2 Human Reason and Consciousness

6.2.1.2.1 Human Reason

In *Descartes' Error*, Damasio says, his primary objective is to propose that "[t]he strategies of human reason probably did not develop ... without the guiding force of ... emotion and feeling" (1994: xii)—and that "certain aspects of the process of emotion and feeling are indispensable for rationality" (1994: xiii). But I think that Damasio has inadvertently revealed something more than he has actually said: he has not chosen to talk about the role played by emotions in human *subjective or phenomenal what-it-is-like/experience* (let alone rudimentary affective awareness), but only the role played by emotions in *human reason*. Rationality is what makes us "distinctly human" (1994: xii). Moreover, Damasio maintains that "[k]nowing about the relevance of feelings in the processes of reason does *not* suggest that reason is less important than feelings, that it should take a backseat to them or that it should be less cultivated" (1994: 246).

6.2.1.2.2 Human Consciousness

'(Human) Consciousness,' for Damasio, is not a thing which allows us to 'feel the biological values of existence' [compare: Sect. 5.3.1.5], but a thing which enables us "to *manipulate* … images, and to apply *reasoning* to them" (2010: chap. 3, sect. 1, my emphasis). He is willing to concede that nonconscious processes are useful insofar as they provide "a suitable and convenient means to execute behavior and give consciousness more time for further analysis and planning" (2010: chap. 11, sect. 2)—but what about cases in which individuals did not enjoy 'consciousness' in the first place? On Damasio's narrow view of 'consciousness,' there would be some individuals (hydranencephalic children, for instance) for whom there would *be* no 'consciousness' for which non-conscious processing could provide that extra time for further analysis and planning. Non-conscious processing would be the only kind of processing these individuals could do.

In which case, I would be interested to see what Damasio thought the point of 'nonconscious processing' in these individuals might be (I must admit, I find it difficult to envision any way of spelling out such an account). The fact that his account of non-conscious processing appears to apply only to individuals in whom non-conscious processing serves to supplement conscious processing detracts from the plausibility of his view. Moreover, the fact that his account of non-conscious processing applies only to otherwise-conscious individuals confirms that the body-map-based view is characterized by its emphasis on human reason and human consciousness. In fact, saying that the body-map-based view *emphasizes* human reason and consciousness may not be strong enough, since one would probably be justified in saying that Damasio focuses on human reason and consciousness *to the exclusion of* non-human forms of 'consciousness' (and/or 'rationality').

This is problematic from the point of view of anybody who is interested in what-it-is-like/experience—because, as I explained [Sect. 4.2], psychological material can influence an individual's what-it-is-like/experience despite being deeply 'unconscious' (in the sense of being unavailable to introspection). Moreover, although we can never be *conscious* of being simultaneously influenced by logically incompatible motivations, such logically incompatible motivations commonly seem to occur [Sect. 4.3], which suggests we are much less 'rational' than we might think.

6.2.1.3 The Cerebral Cortex

Unlike many authors, Damasio believes that "human reason depends on several brain systems, working in concert across many levels of neuronal organization" (1994: xiii), rather than on one specific brain system or area. But at least some potential brain systems are disqualified from being able to contribute to 'consciousness' because on Damasio's view, perception and cognition are central (as are human reason and human consciousness). The third and final point I want to make about Damasio's view is that it is *corticocentric*. This follows logically from my two previous claims, because, as I discussed in Chap. 5, perception, cognition, (introspective) consciousness and reasoning all rely on the functioning of the cerebral cortex. However, even if I couldn't derive 'This view is corticocentric' from 'This is a perceptual-cognitive view, concerning human reason and human consciousness,' Damasio also talks explicitly about his commitments to brain regions.

Several regions of the cerebral cortex "are unequivocally involved in making the very images we behold and manipulate in our minds," says Damasio (2010: chpt. 3, sect. 4). "*Images* [the perception and internal display of which is sufficient for 'mind,' 6.2.1.1] *are based directly on those neural representations, and only those, which are organized topographically and which occur in early sensory cortices*" (1994: 97–8, emphasis in original). What is more, according to Damasio, "those cortices that do not make images tend to be involved in recording them or manipulating them in the process of reasoning, decision, and action" (2010: chap. 3, sect. 4). Therefore, I feel justified in calling Damasio's view 'corticocentric': it is characterized by its emphasis on the functioning of the cerebral cortex.

Damasio does also explicitly discuss the important role of various subcortical brain regions (especially with regard to their role in producing emotions); in fact, this is one of the respects in which Damasio's view stands out among other even more corticocentric views. But he seems to have somewhat grudgingly admitted that this important role is played by sub-cortical structures, only at a fairly advanced point in his career, and only when the strong evidence for the importance of these regions became impossible to ignore (more on this in 6.2.3.3 below). Moreover, as I explained [Sect. 6.1.1.1], he denies that animals 'lower' on the evolutionary ladder than a few of the higher mammals experience proper emotional feelings and/or consciousness (and that the 'highest reaches' of consciousness are unique to humans). And the main biological difference between humans and 'lower' mammals is that humans have a much more highly developed cerebral cortex [Sect. 5.1] (which Damasio considers "the current pinnacle of neural evolution," 2010: chap. 10, sect. 2).

I am not trying to downplay Damasio's important concession to the subcortical brain (especially with regard to feelings)—but even if we are not able to say that Damasio's view *exclusively* concerns cortical processes, what I do think is warranted is saying that cortical processes are *central* to his view. Hence, 'cortico*centric*.' But this is problematic from the point of view of anybody who is interested in what-it-is-like/experience—because, as I explained [Sects. 5.2 and 5.3], there is a rudimentary type of affective awareness which is shared homologously among mammal species, which emerges prior (both in phylogeny and in individual development) to more advanced reflective and/or introspective consciousness; which undergirds reflective/introspective consciousness; and which does not require the cerebral cortex.

6.2.2 Body Maps

On Damasio's view, having a mind is understood to be a question of displaying various types of images internally, and of cognitively ordering and arranging these images (6.1.1.1). One specific type of image which Damasio takes to be particularly central in the construction of 'minds' is what he refers to as a 'body map.'

Even very simple organisms need to be able to detect the presence and/or location of relevant stimuli relating to their bodies so that they can take appropriate survival-enhancing action. When an organism interacts with an external object, sensory information arising throughout the body is used to construct not only neural maps of the relation of the object to the organism, but also neural maps of the organism's own body. What makes body maps different from other sensory representations is that, when engaged in body mapping, the organism simultaneously represents the environmental objects to which it is responding, and represents itself in the act of representing (and responding to) those objects.

Body mapping is unique among other types of mapping because 'the body' (which is "the thing mapped") "never loses contact with the mapping entity" (which is the brain). The main purpose of body maps is to help fine-tune actions; but the actions themselves play a crucial role in creating the maps that inform the actions. Body maps form "a resonating loop that achieves something akin to a body-mind fusion" (2010: chapter. 8, sect. 5). For this reason, body maps also provide the essential foundation for conscious minds: "[t]he entire fabric of the conscious mind is created from the same cloth—*images generated by the brain's map-making abilities*" (2010: chap. 8, sect. 4). It is because (body) map-making abilities form such a central part of Damasio's view that I feel justified in calling it 'the body-map-based view.'

6.2.2.1 The Somatic Marker Hypothesis

Human organisms' proclivity for making body maps helps them during the reasoning process, especially insofar as the ability to make body maps is required to produce what Damasio calls 'somatic markers.'

Sometimes, when faced with a decision, without being able to articulate any of the pros, cons, or relevant factors, we get a strong feeling about which option to prefer—or which option to steer clear of. In Damasio's framework, the term 'gut feeling' is quite apposite, because, in such a case, "the feeling is about the body" (1994: 173). When one considers an undesirable option, one gets an unpleasant somatic feeling—even if that unpleasant feeling only lasts a moment, and even if it feels only very slightly unpleasant. We are all familiar with 'butterflies in the tummy' when we are nervous or excited, or the 'sinking feeling' associated with dread, and this is exactly what Damasio means by 'somatic markers': evaluative assessments of options that are experienced in a *somatic* (bodily) way. They are '(somatic) *markers*' because, on Damasio's terminology, these bodily feelings are associated with (or 'mark') certain 'images' associated with a potential response option.

In Damasio's lexicon, "somatic markers are a special instance of feelings generated from secondary emotions."

> Those emotions and feelings have been connected, by learning, to predicted future outcomes of certain scenarios. When a negative somatic marker is juxtaposed to a particular future outcome the combination functions as an alarm bell. When a positive somatic marker is juxtaposed instead, it becomes a beacon of incentive. (1994: 174)

Somatic markers work on processes of memory, attention or reasoning "so that the decision-making process is biased toward selecting the action most likely to lead to the best possible outcome" (2003: 148–9). They are not 'deliberative' in themselves, but they assist with deliberation by "highlighting some options (either dangerous or favorable), and eliminating them rapidly from subsequent consideration" (1994: 174).

Although Damasio says that somatic markers are a special instance of *feelings generated by secondary emotions*, we know that he understands *feelings* as *perceptions* of the body during an emotional state [Sect. 6.1.1.2]. Somatic markers can themselves therefore be counted as (perceptual) '*images*'—and since "*all* images occur in an aggregate workspace that is formed by separate early sensory regions of the cerebral cortex" (2010:

chap. 8, sect. 4, my emphasis), then logically, somatic markers would also have to occur in these regions.

6.2.2.1.1 Skin Conductance Studies

Empirical data seems to confirm Damasio's theory that the prefrontal cortices are essential for somatic marking. The rationale for skin conductance studies is that any percept, thought or emotion leads to subtle bodily changes: specifically, the skin's sweat glands release a small amount of sweat—not enough to notice by vision or touch, but enough to "produce resistance to the passage of an electrical current" (1994: 207). Experimenters pass a small electrical current through the skin between two electrodes, and record the amount of current that passes between them under different conditions; if the subject feels 'stressed,' one would expect the subject to produce more sweat, which would have the effect of increasing skin conductance. For example, in lie-detector tests, the device measures whether skin conductance is the same when the subject provides an answer that is definitely true, and when the subject provides the 'target' answer. If the conductance is higher during the targeted answer, there is a good chance that the subject is being untruthful (and therefore experiencing stress, and therefore secreting more sweat, which in turn increases skin conductance).

To investigate the SMH, Damasio and his colleagues used skin conductance to compare normal individuals, individuals with lesions outside the [pre]frontal cortices, and individuals who had (ventromedial) [pre]frontal cortical lesions. First, they measured subjects' response to "startling stimuli such as loud noises, or to the behaviors that reliably elicit SCRs ['skin conductance responses'], for example deep breath" (Damasio 1996: 1417). In this condition, they found that the skin conductance was the same for normal subjects, patients with prefrontal lesions, and patients with lesions in other areas. This told the experimenters that there was nothing wrong with the frontal patients' capacity for generating the relevant physiological responses to the stimuli.

For the experimental condition, the subjects had to view a series of slides, while their skin conductance was recorded. Some of the slides showed "emotionally neutral" images, like landscapes or abstract patterns; but others contained 'disturbing' images (for example, "scenes of social catastrophe, or body mutilation," ibid.). What the authors found was that, of their three groups of subjects, the normal subjects and the patients with non-frontal lesions generated plenty of skin conductance in response to the 'disturbing' images, but not the neutral ones. The frontal patients, on the other hand, showed no skin conductance at all, in either of the conditions (Damasio 1994: 209). Not only did the frontal patients show no skin conductance; they also reported verbally that they experienced no particular feelings when viewing these images. This told the experimenters that these patients no longer had "a normal ability to generate somatic responses to stimuli with an emotional component" (Damasio 1996: 1417).

Given that the control subjects showed elevated skin conductance when viewing the disturbing images, while the frontal patients showed none, Damasio's conclusion (that the frontal patients experienced little or no 'stress' while viewing the 'disturbing' images) seems reasonable. Both their skin conductance and their verbal reports were in agreement on this. But what I would like to know is what it is, exactly, that allows Damasio to claim that these results support his *somatic marker hypothesis*, in particular.

Even if we accept, for the sake of argument, that the frontal patients did lack the normal mechanism for producing emotions on viewing the disturbing images, it does not seem to me as if the skin conductance studies have the power to shed much light on whether or not this mechanism is anything like Damasio's 'mechanism for somatic marking.' Skin conductance is a measure of 'stress' in a very broad sense, in the sense not only of 'anxiety' or 'tension,' but also of 'vigilance' or 'alertness.' For instance, if test subjects were required to perform any kind of quite difficult test, and if they could predict exactly when the critical stimulus was going to appear, their skin conductance levels might be expected to rise just before that stimulus appeared. The subjects' whole bodies would respond with a flood of adrenalin, perspiration, increased pulse, et cetera, in order to be as well-prepared as possible at the critical moment. Similarly, if suspected criminals were attempting to fool their interrogators into believing their (false) stories about their own innocence, lie detector equipment would be expected to show elevated skin conductance as they gave those untruthful (stressful) answers.

Therefore, I cannot see any reason to assume, as Damasio does, that the capacity for experiencing the appropriate emotional feeling in a disturbing situation necessarily requires a capacity for experiencing 'somatic markers'—any more than the capacity for experiencing somatic markers is required in order to feel 'stressed' while being untruthful. Accordingly: although these results are quite *compatible* with the SMH, I would also think they fell rather far short of actually *supporting* the SMH.

6.2.2.1.2 PET Studies

PET (positron emission tomography) studies also provided Damasio with some of his evidence for the SMH. For this technique, short-lived radioactive (positron-emitting) molecules, especially certain types of sugars, are administered intravenously (Panksepp 1998a: 90–1). By monitoring the movement of these molecules around the brain, investigators are able to visualize which brain regions are consuming the most sugar—which is to say, which regions are the most active (Damasio 2003: 97–98).

Damasio tested the hypothesis that the occurrence of feelings depends on "significant engagement of the areas of the brain that … map the ongoing state of the organism" (Damasio 2003: 96)—brain areas which include both cortical and subcortical structures. He instructed the participants each to think of an emotional episode in their lives; they were then to think about those episodes in as much detail as possible, and call to mind as much mental imagery as possible, while PET technology was used to determine which brain areas were most active. Damasio found that "all the body-sensing areas under scrutiny … showed a statistically significant pattern of activation or deactivation"—which he took to indicate that "the mapping of body states had been significantly modified during the process of feeling" (2003: 99–100).

However, according to Panksepp,

> [o]ne of the shortcomings of [different neuroimaging techniques. including PET] has been the relatively long time window for observing brain

> activities. One has to harvest data for a minimum of ... about 10 seconds, which is still much too long to follow the flow of ongoing cognitive activities in real time. (1998a: 91)

This means that there is some doubt about whether it is even theoretically possible for PET technology to show a correlation between 'the process of feeling' and 'the mapping of body states,' as Damasio claimed. Moreover, even if the brain regions highlighted during Damasio's study did include 'body sensing regions,' as he claimed, 'body sensing' is by no means the only thing these regions are responsible for. For instance, they are also known to be important for "the initiation and organization of complex behavior" (Panksepp 2003a: 124)—which means that the activation of these regions might easily have to do with the 'executive' or motivational aspects of the subjects' *behavior*, rather than the 'body mapping' (or somatic marking) aspects of their (emotional) *feeling*.

This being the case, it seems that, even if these PET results are *compatible* with the somatic marker hypothesis, they don't necessarily *support* it all that much. There is general agreement among neuroscientists that the somatosensory cortex is central to producing *sensory* 'feelings,' but "there is little evidence that ... *bodily indices* of emotions *precede* and *cause* emotional feelings" (Panksepp 2007a: 102)—which is what the SMH entails. Therefore, according to Panksepp, Damasio's somatic marker hypothesis "remains a weak working hypothesis."

In any case, whether or not the SMH is plausible in itself, we can clearly see that it is characterized by those five main commitments of the body-map-based view. First: body maps (which are needed to enable somatic marking) are constructed from *perceptual* information about the body [Sect. 6.2.2]; and second, feelings (including somatic markers) are *cognitive* phenomena [Sect. 6.2.1.1]. Thirdly, the point of somatic markers is to act as alarm bells or beacons of incentive for people who are deliberating, and they are presumably not very much to the purpose in the case of individuals who are not capable of (*rational*) deliberation in the first place [Sect. 6.2.2.1]. Fourthly, somatic markers, like other feelings, only have their full effect when one is able to 'know' those feelings: when one is not only 'minded,' but also '*conscious*' [Sect. 6.1.1.1]. And finally: even if Damasio had not specifically stated that somatic markers rely on cortical functioning [Sect. 6.2.1.1], we already know that almost everything to do with perception, cognition, consciousness, or rationality requires the *cerebral cortex* [Sects. 5.1 and 5.2].

As I have explained in Chaps. 3, 4, and 5, this is problematic for anyone who is interested in what-it-is-like/experience.

6.2.2.1.3 Phineas Gage

The most famous clinical case from which Damasio draws some of his evidence for the SMH is undoubtedly the case of Phineas Gage (Cacioppo and Berntson 2004: 977). In 1848, Gage was working as a construction foreman in New England, laying down the tracks for a new railway. He was seen by his superiors as efficient and capable; just the right kind of man to be in charge of a project requiring both physical strength and mental acuity (Damasio 1994: 4). But one day when he was momentarily distracted, Gage accidentally struck his tamping iron directly onto the charge of explosive being used to blast the rock, *without waiting for his colleague to cover the charge with a layer of sand*—and he became famous for not only driving an iron bar through his own head, but also living to tell the tale. In fact, Gage not only survived the accident, but showed very little sign that he had had any accident at all. His limbs convulsed a little when he landed (after being thrown several feet across the ground) following the blast, but he was able to speak after a very few minutes. When he had been transported to the doctor in an ox cart, he was even able to climb out of the cart unassisted.

In less than two months, the doctors pronounced Gage fully recovered. But physician Doctor Harlow reported that the "balance ... between his intellectual faculty and animal propensities" had been disrupted. Although he was "[a] child in his intellectual capacity and manifestations," he nevertheless had "the animal passions of a strong man" (in Damasio 1994: 8). Gage had previously been considered 'temperate,' 'well-balanced' and 'persistent'—but now he started to use profanity in his speech, became impatient and impulsive, and repeatedly made plans for the future which he subsequently abandoned. His friends and family had trouble recognizing the old Phineas in this post-trauma individual, and his employers decided that they were no longer able to employ him, on account of his 'personality changes.'

Although medical techniques were quite limited at the time of Gage's accident, his skull was preserved at Harvard medical school—allowing Hanna Damasio to estimate which areas of the brain must have been damaged, given the locations on the skull where the tamping iron penetrated. Drawing from her findings, (Antonio) Damasio asserts that damage to the 'ventromedial' aspect of the prefrontal cortex

> compromises, in as pure a fashion as one is likely to find, both reasoning/decision-making and emotion/feeling, especially in the personal and social domain. One might say, metaphorically, that reason and emotion 'intersect' in the ventromedial prefrontal cortices. (1994: 70)

On Damasio's view, Gage experienced problems relating to "the ability to anticipate the future and plan accordingly within a complex social environment; the sense of responsibility toward the self and others; and the ability to orchestrate one's survival deliberately at the command of one's free will" (1994: 10). Damasio takes it to be self-evident that (1) this is a reasoning defect, and that (2) such a defect cannot be ascribed to purely cognitive factors (seeing that Gage's other cognitive skills seemed intact). However, the reasoning defect *can* be accounted for, in Damasio's opinion, by saying that Gage, post trauma, lacked the necessary mechanisms for somatic marker signalling. Gage, like the other frontal patients Damasio examined, could not "generate emotions relative to the images conjured up by certain categories of situation and stimuli, and thus [could not] have the ensuing feeling" (1994: 138). In the kinds of situations in which somatic markers would normally have been functioning as alarm bells and beacons of incentive [Sect. 6.2.2.1], Gage had nothing—and this was the root of all his problems.

But although it is clear that *something* is going wrong with some of Gage's abilities, it is much less clear to me that what is going wrong has anything to do with 'somatic markers.' Damasio has identified certain problems that Gage had relating to social conventions, and ethical rules; and to planning future actions within a complex social environment—but I would say that these problems can all still comfortably be ascribed to a *cognitive* deficit. 'The ability to anticipate the future and plan accordingly' is primarily a cognitive ability, whether or not the anticipating and planning take place 'within a complex *social* environment.' Similarly, 'the ability to orchestrate one's survival' is primarily a cognitive ability, whether or not this orchestrating is being done 'deliberately,' by one's own 'free will.' Although it might be reasonable to say that these particular cognitive activities involve 'social' cognitions, the fact that these cognitions are 'social' in no way changes the fact that they are *cognitions.* Since we already know that the pre-frontal cortex is crucial for all cognitive and 'executive function' tasks [Sect. 5.1]—like processing social conventions, planning future actions within a complex social environment and orchestrating one's survival—it is not very surprising that Gage had trouble in these areas.

Moreover, none of this seems like any kind of good reason for suggesting that Gage's deficits have anything to do with his 'emotions.' Damasio himself has already admitted that Gage, after his accident, had 'the animal passions of a strong man'—which suggests that his primary emotions were healthy. And if he had a problem with his secondary emotions (in the form of somatic markers) only, in spite of his healthy primary emotions, then it makes sense to assume that this 'emotional' problem was due to a *cognitive* deficit—seeing that, as I explained [Sect. 6.1.1.2], the important *cognitive* element of secondary emotions is what differentiates them from primary emotions.

6.2.2.2 The Body-Map-Based View on the Evolutionary Advantages of Consciousness: Knowing Your Feelings

Philosophers (such as Chalmers 1996) sometimes claim that "phenomenal consciousness *cannot* be adaptive because it has no specific function" (Montemayor and Haladjian 2015: 177). But proponents of biological naturalism (such as Damasio, Panksepp and I) take it for granted that all the traits which can be observed in organisms must, by definition, have provided those organisms (or the forerunners of those organisms) with some kind of adaptive (evolutionary) advantage—otherwise, those traits would have not have been preserved.

'Mind' (in Damasio's sense of having the capacity to display and order images internally) is not necessary for "balancing the internal milieu" (Damasio 1999: 303), and the early forerunners of our higher mammal species managed perfectly well without minds: they "operated on the basis of dispositions" (2010: chap. 6, sect. 3). The capacity for 'mind' (that is, the capacity to display images/maps internally) gave 'minded' creatures a survival advantage over their un-minded counterparts, insofar as it enabled them to move beyond simple dispositional responses. When creatures achieve 'consciousness' (in Damasio's sense), they become capable of "more deliberated regulation, based on a mind equipped with identity and personhood and now actively seeking not mere survival but certain ranges of well-being" (2010: chap. 2, sect. 10). It therefore comes as no surprise to find that the body-map-based view of the evolutionary advantages of consciousness is characterized by the same perception/cognition/consciousness/rationality/cortex bias which I detected in so much of Damasio's work, and which I thought detracted from the plausibility of his view, as I will now explain.

6.2.2.2.1 Novel/Flexible Responses

The first evolutionary advantage of 'consciousness,' according to the body-map-based view, is that being conscious allows organisms to produce novel responses to new situations. Primitive organisms can only react to stimuli in a reflexive or stereotypical way (for instance, cubomedusas are able to respond to stimuli by moving in one of four basic directions, 5.3.1); but being 'conscious' enables more complex organisms to produce appropriate, multi-layered responses even in situations they have never before encountered (ibid.).

This capacity for producing novel responses also gives 'conscious' organisms much more behavioral *flexibility*. They not only have the capacity to learn from experience, but also the capacity to pick up on quite subtle (environmental or social) cues, and fine-tune their behavior accordingly (Damasio 1994). Conscious creatures are able "to *adapt* to a large number of niches and are able to *learn to adapt* to an even greater number" (2010: chap. 2, sect. 10).

Because learning, adapting, and planning are largely cognitive capacities, involving reason (and therefore relying on the cerebral cortex), what Damasio has to say about the evolutionary advantages of (human) 'consciousness' reinforces my claim that his view is characterized by an emphasis on perception/cognition, (human) reason/consciousness, and/or/therefore the functioning of the cerebral cortex.

6.2.2.2.2 Planning Ahead

Secondly, Damasio suggests that 'consciousness' is advantageous to organisms insofar as it provides complex organisms with a way to anticipate their needs before these needs become critical—which is to say, to seek out whatever they need *proactively*, rather than just *reacting* by the time the need has become very urgent. In 'conscious' creatures, the (mental) images common to all 'minded' creatures can, in addition, be "guided by reflection and used for *effective anticipation of situations, previewing of possible outcomes, navigation of the possible future,* and *invention of management solutions*" (2010: chap. 7, sect. 7).

It seems to me that this kind of need-anticipation can justifiably be described as a cognitive advantage, mostly useful to rational creatures, and especially if they are (introspectively) 'conscious.'

6.2.2.2.3 Feelings/Incentive

For Damasio, the third—and most important—advantage of consciousness is that it allows conscious organisms to be introspectively aware of their emotional feelings [Sect. 6.1.1.2]. Some situations the organism may encounter "may be *so* potentially advantageous or *so* potentially detrimental to its survival" that mere 'emotions' (that is, 'largely automated programs of actions,') are not sufficient to ensure that the organism deals with the situation successfully (for example, 2010: chap. 2, sect. 8). The additional 'incentive' provided by conscious feelings would be required before the organism would be sufficiently motivated to take the necessary action. But *knowing* that one is having a feeling—that is, being 'conscious' of feelings—provides even more of an advantage that just feelings alone, because *knowing* one's feelings not only provides incentive, but also facilitates the planning of novel and/or flexible responses.

In other words, for Damasio, being 'conscious' has the first two specific advantages I identified; but being conscious *of feelings*, in particular, has both these advantages, as well as the additional advantage of providing 'incentive.' To the extent that, on the body-map-based view, the evolutionary advantages of (introspective) human consciousness and/or human reason are conceptualized in terms of perception, cognition and the functioning of the cerebral cortex—and to the extent to which, as I have explained in the preceding chapters, such a conceptualization is problematic—the body-map-based account of the evolutionary advantages of consciousness is problematic.

6.2.2.3 The Lesson (of 'Body Maps') for Philosophy

I have already explained how Damasio's concern with perception/cognition, and with human reason/consciousness, led him to quite a corticocentric position overall [Sect. 6.2.1]. In addition, what we have now seen is that Damasio's somatic marker hypothesis, as well as his account of the evolutionary advantages of consciousness, exhibit these same hallmarks.

Any view which assigns a central role to body maps and/or somatic markers must be characterized by its emphasis on perception, cognition, human reason and human consciousness, as well (therefore) as by its emphasis on the functioning of the cerebral cortex; as must any view which understands the evolutionary advantages of consciousness—of knowing your feelings—in similar terms. (Introspective) *consciousness* of feelings (understood in terms of the *perception/cognition* of images, and enabled by the cerebral *cortex*) draws the organism's attention towards the potential consequences of situations—and thus make it possible for all the organism's experiences "to influence *the reasoning and decision-making process*" (2003: 178, my emphasis) to the maximum.

6.2.3 Damasio's Views on the Subcortical Brain

> Books should be written for a reason, and this one was written to start over. I have been studying the human mind and brain for more than thirty years, and I have previously written about consciousness in scientific articles and books. But I have grown dissatisfied with my account of the problem. (2010: chpt. 1, sect. 1)

One of the specific regards in which Damasio's views have changed over time is how he envisions the role of the subcortical brain.

6.2.3.1 Early Views

6.2.3.1.1 Descartes' Error (1994), Brainstem and Life Regulation

The discrepancy between the kinds of thing the cortex does and the kinds of thing the subcortical brain does appears so blatant that it "fostered [the] implicit and seemingly sensible view [that] [t]he old brain core handles basic biological regulation[5] [and] the neocortex deliberates with wisdom and subtlety" (1994: 128). But actually, Damasio claims (in *Descartes' Error*), this is not an accurate view of the division of labor at all:

> Nature appears to have built the apparatus of rationality not just on top of the apparatus of biological regulation, but also *from* it and *with* it. (1994: 128)

In other words, Damasio appears to believe that the 'dumb arousal' view [Sect. 5.3.1.3] is problematic. However, even if he admits that the (cortical) 'apparatus of rationality' is built *from* and *with* the biological regulation functions of the brainstem (rather than just *alongside* brainstem functions), Damasio does not appear to be ascribing to the brainstem itself a significant role in the actual building of that apparatus of rationality. He may be conceding that the (cortical) apparatus of rationality is built *from* and *with* the (brainstem) apparatus of biological regulation—but he is certainly not conceding, for instance, that the apparatus of *biological regulation* is built from or with the apparatus of *rationality*. He seems to be quite definite about which brain structure *enables the cortex* to produce rationality, as opposed to which brain structure is responsible for actually *producing* the rationality. In his earliest work, therefore, we would probably be justified in saying that Damasio subscribes to something like the 'dumb arousal' view of brainstem function.

6.2.3.1.2 The Feeling of What Happens (1999), Brainstem and Emotions

In *The Feeling of What Happens*, Damasio to some extent moves away from his previous focus on 'human reason,' and focuses more on human emotions and feelings. In a sense, he suggests, the projects and goals that humans pursue can all be taken as aiming for the same thing: the achievement of a positive emotional state, and the avoidance of negative emotions (1999: 35). This shift in Damasio's focus entails a corresponding shift towards more emphasis on subcortical brain circuits: he concedes that "[t]he devices which produce emotions occupy a fairly restricted ensemble of subcortical regions" (1999: 51).

However, I am claiming that quite a strong corticocentric commitment can still be detected even here. During the production of emotions, Damasio says, "[f]irst, the brain *induces* emotions *from* a remarkably small number of [subcortical] brain sites." Then, the (cortical) 'induction sites' "are involved in processing" the emotions. Finally, "some of these [cortical] sites are also involved in the recognition of stimuli which signify certain emotions" (1999: 60, my emphasis). I said (above section) that even Damasio's concession that the (cortical) apparatus of rationality was built *from* and *with* the (subcortical) apparatus of life regulation was quite a far cry from saying that the apparatus of life regulation was built from and with the apparatus of rationality. Similarly, here, close examination of Damasio's sentence reveals that it is still the cortex which is playing the important role: it is *inducing* the emotions *from* the subcortical sites—and evidently, if there was no cortical induction involved in the process, the subcortical sites on their own would not be able to generate emotions.

Perhaps it would be unfair to say that Damasio (in *The Feeling Of What Happens*) is taking a 'dumb arousal' view of brainstem function: insofar as he is interested in the brainstem's role in the production of *emotions*, perhaps it would be unfair to characterize this (1999) view as being a view of 'arousal' *at all*. However, even though he has acknowledged that some of the steps in the process of experiencing emotions are subcortical, he also maintains that the cortex is necessary for other steps: therefore, it would probably still be fair to call his 1999 view 'corticocentric.'

6.2.3.2 Intermediate View

6.2.3.2.1 Looking for Spinoza (2003), Brainstem and Feelings/Body Maps

In *Looking for Spinoza*, Damasio starts to assign an even more central role to feelings. However, at the same time as he is starting to take the role of *feelings* more seriously, to some extent he also retreats from the more charitable view of the *brainstem* he set out in *The Feeling of What Happens* (Panksepp 2003a: 125). He says that experiencing an emotion comprises "a complicated chain of events" which begins with the "appraisal and definition of an emotionally competent stimulus" (Damasio 2003: 64). Not surprisingly, this first step takes place in one of the brain regions dedicated to sensory processing (which are mostly in the cerebral cortex, 5.2). Next, "signals related to the presence of that stimulus are made available to a number of emotion-triggering sites elsewhere in the brain." For example, in the case of fear, the presence of the stimulus is conveyed to the well-known emotion-triggering site in the (subcortical) amygdala. The final step in the process is that "[t]he emotion-triggering sites subsequently activate a number of emotion-execution sites elsewhere in the brain" (2003: 58)—including 'subcortical' regions (2003: 63). When these steps have been completed, the result is the "full-blown emotional state" (2003: 64). However, although Damasio has specified the steps of the process involved in the production of emotion *in rather more detail* here than he did in *The Feeling of What Happens* [Sect. 6.2.3.1], close examination of these details seems to indicate that the actual role he ascribes to the subcortical brain in the production of emotion is still much more of a supporting role than a leading role.

'Feeling,' for Damasio, is a question of the perception of body states; therefore, in the same way we would not be able to see if the visual regions of the brain were damaged, so "[i]f those body-sensing regions [which are predominantly in the cerebral cortex] were not available to us we would not feel anything" (2003: 111). *Knowing* one's feelings is a question of being 'conscious' of feeling those feelings—which requires a sense of self, and therefore relies on the function of the cerebral cortex [Sect. 5.1.1.2]; therefore, the implication seems to be that if 'those *consciousness*-enabling regions [predominantly in the cerebral cortex] were not available to us,' we would equally not know that we feel anything.

Moreover, Damasio draws an analogy between 'the maps in the brain of a living organism' and 'the maps in the cockpit of the Boeing 777' which I think is highly revealing. He says that the 'monitoring devices' in airplanes "are but a pale version of the monitoring devices in the central nervous system of a complex living organism" (2003: 127). Since even the monitoring devices of the Boeing have very impressive monitoring capabilities, the fact that Damasio thinks these devices are inferior to the brainstem shows that he is no longer denying the impressive capabilities of the brainstem. However, the fact that he considers the brainstem body mapping apparatus to be *in any way* similar to aircraft navigational systems reminds us that he has a dumb-(or smart-)arousal-type view of brainstem function.

In the case of the aeroplane, the sophisticated cockpit navigational systems may well play a vital role in successfully flying the plane, but, were it not for the engines which were responsible for getting the plane off the ground in the first place, there would be no need for any navigational equipment to do anything at all. Similarly, even though Damasio admits that the 'monitoring and mapping' functions of the brainstem (like the plane's navigational equipment) are vital to the functioning of the organism, he does not appear to be acknowledging that the brainstem might also have a much more fundamental role to play—that it could be more analogous to the engines that actually get the plane off the ground than merely the 'navigational systems' that stop the plane crashing.[6]

Therefore, we would probably still be justified in concluding that, on Damasio's intermediate view, the brainstem is playing some kind of supportive role towards the cortex, which is what does the really important work (hence, 'the view is corticocentric').

6.2.3.3 Mature View

6.2.3.3.1 Brainstem and Arousal/Life Regulation

I claimed that, especially in his earlier work, Damasio seemed to see the brainstem as mainly responsible for some basic 'life-regulation' functions, as well as for 'switching on' the cortex [Sect. 6.2.3.1]. In 2010, Damasio admits that the dumb arousal view "does not do justice to the results [brainstem nuclei] achieve" (2010: chap. 10, sect. 2). He admits that nuclei in the upper brainstem not only are essential to 'life management,' but also constitute "the neural home of biological value," which is, in all likelihood, "the place where the process of making mind begins, in the form of primordial feelings" (ibid.). However, even if he now thinks the brainstem's role is more like 'smart arousal' than 'dumb arousal,' I think it is still fair to say that he sees the brainstem as playing some kind of 'arousing the cortex' role, rather than an important role of its own. This will become clearer once I have contrasted the body-map-based view with the affect-centric view on the subcortical brain [Sect. 6.3.3 (see also Sect. 6.3.2)].

6.2.3.3.2 Brainstem and Body Maps

In 2003, Damasio still viewed the role of the subcortical brain as some kind of a 'monitoring' role [Sect. 6.2.3.2]; even if he said the monitoring devices in the brainstem were vastly superior to the monitoring devices in the Boeing, he was still not conceding that brainstem structures had an active role of their own, as opposed to just *enabling the cortex* to have a role.

His 2010 view is that "the critical body-mapping and image-making structures are located below the level of the cerebral cortex, in a region known as the upper brain stem" (2010: chap. 1, sect. 7)—so, seeing that we know how important body maps are to Damasio, it sounds as if conceding that the brainstem is critical to body mapping would be quite a significant concession for him to make. However, he maintains that, while upper brainstem structures are "able to create coarse maps," it is only the cerebral cortices which really "excel at the creation of detailed maps" (2010: chap. 3, sect. 2). Therefore, conceding that structures in the upper brainstem are critical for body-mapping and image-making may not be all that much of a concession to make—and Damasio could probably still be described as having a corticocentric view.

6.2.3.3.3 Brainstem and Feelings

Even in 2003, Damasio admitted that some of the steps of the emotion process were subcortical [Sect. 6.2.3.2]. In 2010, he concedes that most basic feelings "originate at the level of the brain stem rather than the cerebral cortex" (2010: chap. 1, sect. 7). However, it seems as if he is allowing that they originate at this level only insofar as this level of brain organization facilitates the creation of body maps; he is certainly not conceding that it could be, for instance, *the rudimentary affective awareness* produced by the brainstem which is essential for feelings. And I have just explained why conceding that the brainstem may be essential for creating coarse maps may not be much of a concession to make. Accordingly: conceding that basic feelings originate at the level of the brainstem is probably not much of a concession to make—and it would probably still be fair to describe Damasio's view as 'corticocentric.'

I said [Sect. 6.2.3.2] that Damasio seemed to come very close to saying that 'if the consciousness-enabling regions [in the cerebral cortex] were not available to us, we would not feel anything.' But if this were the case, then presumably hydranencephalic children (some of whom lack both the 'body-sensing regions' *and* the 'consciousness-enabling regions' of the cortex) would not feel anything. This is patently not the case, however, as Damasio himself acknowledges: hydranencephalic children are capable of sentience, feeling and emotion. He attempts to get himself out of this difficulty by saying that "[t]he degree of sentience, feeling, and emotion that is possible in these cases is quite limited, of course, and, most important, disconnected from the wider world of mind that, indeed, only the cerebral cortex can provide" (2010: chap. 3, sect. 5)—but in my opinion (as in Panksepp's), this attempt does not succeed. Damasio seems to be doing little more, here, than paying lip-service to a proposition he is unable to deny completely, but does not really believe in. He is acknowledging that the cortex is not essential for *absolutely all* sentience, feeling and emotion, while maintaining that subcortical structures are only capable of producing a *limited degree of* sentience, feeling and emotion.

According to Panksepp, Damasio's original view (for example, in *Descartes' Error* and/or *The Feeling of What Happens*) was that emotional feelings were primarily produced in the somatosensory cortex; and only as a result of "a flood of brain imaging evidence" did not only Damasio (in *Looking for Spinoza)*, but virtually everybody else as well, come round to the view that more primitive (subcortical) brain regions are more important for "certain affective experiences" (Panksepp 2003a: 115). Panksepp seems to be suggesting that, had this flood of brain imaging evidence not left Damasio with *no alternative* to moderating his views slightly, Damasio probably *would not have* moderated his views—and as it is, he moderated them only very slightly.

6.2.3.3.4 Brainstem and Conscious Minds

In *Self Comes to Mind*, Damasio concedes that "[b]rains begin building conscious minds not at the level of the cerebral cortex but rather at the level of the brain stem" (2010: chap. 1, sect. 1)—and since humans are (usually) 'conscious' in the 'extended' (self-aware) sense, it sounds as if conceding that the brainstem is important for building conscious minds would be quite a significant concession. But as it turns out, Damasio thinks that the brainstem is essential for conscious minds only *insofar as* it is (also) essential for body maps; and as I have already explained, he thinks that the brainstem is only capable of producing 'coarse maps' [Sect. 6.2.3.3], whereas the cerebral cortex excels at the production of detailed maps.

As I mentioned [Sect. 5.1.1.2], several studies in the mid-1900s showed that lesions in certain upper brainstem sites resulted in somnolence or even coma (Damasio mentions the study by Plum and Posner). To Damasio, these findings seemed to confirm his view regarding the centrality of body mapping: "only above the level of the midpons [in the brainstem] does the collecting of *whole-body information* become complete"—which is why "[o]nly above this level does the brain possess all the information it needs to create comprehensive maps of the whole body" (2010: chap. 10, sect. 2). Therefore, the fact that brain damage at this level leads to impairments of consciousness seemed to confirm Damasio's theory that the brainstem was critical for the production of conscious minds—*insofar as the brainstem was also critical for the production of body maps* (see above, 'Brainstem and feelings'). The thing is, even if Damasio admits that consciousness begins in the brainstem, insofar as he takes consciousness to be based on body maps, and insofar as he claims it is only the cerebral cortices which 'excel at the creation of detailed maps' (above, 'brainstem and body maps'), this is still problematic.

Presumably the type of 'consciousness' which 'begins' in the brainstem would be quite a lot inferior to the 'extended' consciousness [Sect. 6.1.1.1] which requires the cortex. In which case, this 'concession' about the brainstem's role in consciousness may not really be very much of a concession at all. Moreover, seeing that Damasio *defines* 'feelings' as 'conscious' (that is, only available to creatures who have both 'mind' and 'self,' Sect. 6.1.1.2), admitting that the brainstem is essential for conscious minds is no different from admitting that the brainstem is essential for feelings—and I have already explained why Damasio's admission that the brainstem is essential for feelings is problematic [Sect. 6.2.3.3.3].

6.2.3.4 The Lesson (of 'Damasio's views on the subcortical brain') for Philosophy

As I explained [Sect. 6.2.1.3], Damasio thinks that the cerebral cortex represents 'the pinnacle of human evolution,' and is essential not only for 'consciousness,' but also for most of the other important capacities he discusses. The brainstem is where the rudimentary images, which are the essential components of 'body maps' (as well as the rudimentary images which constitute feelings), start to take shape; but "external objects exist as images only in the cerebral cortex and cannot be fully imaged in the brain stem."

> The consequence of this economic division of roles is a fatal and complete interdependence of brain stem and cortex. They are *forced* to cooperate with each other. (2010: chap. 10, sect. 3)

In other words, despite his grudging concessions to the role of the brainstem with regard to body maps, feelings, and even conscious minds, Damasio still sees subcortical brain structures essentially as having some kind of 'supporting the cortex' role. Exactly how the subcortical structures perform that role is neither very important nor very interesting, for Damasio—as long as they do whatever is necessary to enable the cortex to perform its (much more important and interesting) functions. Therefore, to the extent that Damasio essentially still takes a kind of a '(dumb/smart) arousal' view of brainstem function, the body-map-based view is corticocentric—which, I argued in Chap. 5, is problematic.

6.2.4 The Body-Map-Based View: Conclusion

In this Sect. 6.2, I have applied my critical attitude towards perceptual/cognitive/corticocentric views (of human consciousness and/or human reason) to a specific example of one such view.

Damasio himself states explicitly that he is interested in 'human consciousness' and 'human reason'; in addition, I claimed that, because he said that mind was comprised of images, his view was characterized by its emphasis on perception. Because he said that 'consciousness' (and even *perception itself!*) was a cognitive phenomenon, I claimed that the body-map-based view was also characterized by an emphasis on cognition. And because, as I explained [Sect. 5.1], most perception/cognition and consciousness/rationality requires the cerebral cortex, these first two characteristics of the view entailed that it was also corticocentric. Damasio also explicitly states the respects in which his view emphasizes the cerebral cortex.

Body mapping, for Damasio, is of central importance not only to maintaining biological life, or to navigating the environment successfully, but also to mind, feelings, and ultimately, consciousness itself. I claimed that even the somatic marker hypothesis—which is ostensibly a hypothesis about the function of 'emotions'—could be described as showing the same perceptual-cognitive (and therefore corticocentric) commitments, especially as they apply to human reason and consciousness. Even the body-map-based view of the evolutionary advantages of consciousness is characterized by these same commitments.

Even Damasio's more recent view, which concedes that brainstem function is required for body maps and/or feelings and/or conscious minds, is still essentially a glorified 'dumb arousal' view: although he concedes that the brainstem plays a crucial role in these respects, he apparently is not ready to accept that the role of the brainstem goes much beyond *enabling the cerebral cortex* to perform the actual functions.

Damasio, who is not a philosopher, is primarily interested in finding out whether the available empirical evidence supports his hypotheses. But I am not too concerned with whether these hypotheses are true or false. I am more concerned about the fact that, because Damasio's (biological naturalist) body-map-based view is characterized by its emphasis on perception, cognition, human reason and human consciousness (and therefore the functioning of the cerebral cortex), nothing about this view is much use *to anyone who is interested in what-it-is-like/experience.*[7]

6.3 Jaak Panksepp: An 'Affect-Centric' View of Primary-Process Consciousness

With his 1998 textbook of the same name, Panksepp coined the term 'Affective Neuroscience' to refer to the discipline which focuses on "those neural processes that undergird our emotional experiences and actions [and are] essential mechanisms for the generation of the basic emotional forces that we still share with other mammals" (1998a: 14). At that time,

Panksepp said, there was no umbrella discipline synthesizing behavioral, psychological, and neurological perspectives—but he believed that AN could become such a discipline. Such a 'new perspective' as he envisioned would be able to contribute meaningfully to philosophical questions of 'the mind' and 'the body'; he even remarked that he looked forward to the day when "*neurophilosophy* (as heralded in a book by that name written by Pat Churchland...)" would become "an experimental discipline that may shed new light" (1998a: 5) on the 'problem' of the 'explanatory gap.' According to Mortimor Ostow, the first two 'revolutions' in twenty-first century psychiatry were psychoanalytic psychiatry and psychopharmacologic psychiatry; and *Affective Neuroscience* is fast becoming "one of the scriptures of the third revolution" (2004: xvi).

For the purpose of contrasting it with Damasio's 'body-map-based' view, I will refer to Panksepp's view as 'affect-centric,'[8] because the claim that "emotional processes, including extremely subjective feelings ... play a key role in the causal chain of events that control the actions of both humans and [other] animals" (Panksepp 1998a: 14) is its most central tenet. In **6.3**, I will explain how the affect-centric view of the mind-body-self, unlike the body-map-based view, is *not* characterized by its emphasis on (1) perception or cognition, (2) (human) reason or consciousness, or (3) the functioning of the cerebral cortex; and should therefore be preferred by anyone who is interested in what-it-is-like/experience.

Just as in the case of my discussion of Damasio's views [Sect. 6.2], I will not be attempting to give a comprehensive review of everything Panksepp thinks; rather, I will try to bring out the contrast between some specific commitments of Panksepp's affect-centric position and the corresponding commitments of Damasio's body-map-based view.

6.3.1 Disagreeing with Damasio

According to Damasio, Panksepp "links body and self, by means of an innate representation of the body in brain stem," as does Damasio himself (although "[Panksepp's] view of how such a representation contributes to consciousness is entirely different from [Damasio's]," Damasio 1999: 155). Evidently, Damasio means to imply that, insofar as Panksepp's view differs from his own, Panksepp's view is problematic. On the other hand, Damasio's commitment to the idea that "most mental states consist of various types of bodily awareness," according to Panksepp, "is bound to remain a controversial way to conceptualize the activities of the higher reaches of the human mind" (Panksepp 2003a: 115). He thinks that "our thoughts and ideas are as much about events in the world as states of our body" (Panksepp 2003a: 127). In other words, Panksepp thinks that Damasio's view is problematic.

It was probably inevitable that these two authors would each take the other's view to be problematic; however, as I will explain in this section, I favour Panksepp's view (since he and I seem to share similar concerns about the central commitments of the body-map-based view). The affect-centric view of the mind-body-self, in contrast to the body-map-based view, is *not* characterized by its emphasis on:

6.3.1.1 Perception Or Cognition

6.3.1.1.1 Perception

Damasio said that emotions were 'actions accompanied by ideas and certain modes of thinking,' while emotional *feelings* were 'mostly *perceptions* of what our bodies do during the emoting' [Sect. 6.1.1.2]. Panksepp notes that "early investigators, such as William James, suggested … that 'the bodily changes follow directly the PERCEPTION of the exciting fact and … our feeling of the same changes as they occur is the emotion'" (1998a: 33)—which sounds almost identical to Damasio's general position (and in fact, Panksepp refers to the type of view Damasio takes as a 'James-Lange' view).

But Panksepp does not approve of this:

> By remaining so sensory-centric, Damasio may be advancing an incomplete and potentially incorrect theoretical view of *emotional* feelings. Although emotional arousal obviously generates feedbacks that arise from peripheral autonomic and somatic changes, to see these as the major causal

> components of *emotional* feelings may be the smaller, and perhaps more obvious, part of the mystery that needs to be revealed. (2003a: 123)

Clinical cases also cast doubt on any view, like Damasio's, according to which 'feelings' are a kind of bodily perception. For example, spinal cord injury patients lack virtually all ability to sense anything about their bodies below the level of the injury; yet they usually display fairly typical emotional responsivity (Panksepp 1998a: 32). In fact, we have already seen [Sects. 5.2 and 5.3] that people who lack most or all perceptual abilities, owing to *their lack of a cerebral cortex*, display fairly typical—or even exaggerated—emotional reactions.

On Panksepp's view, "strong emotions ... most certainly are *not* just another type of qualia on the same footing as the redness of red" (2003a: 129), as they appear to be on Damasio's view.

> Emotions are NOT just the 'feeling of what happens' [that is, Damasio's 1999 title], but more 'by God, I am going to make this happen!!!' and experiencing such action urges are at least as important to the core of feeling as the sensory feedback about autonomic and visceral changes. Panksepp (2003a: 122)

6.3.1.1.2 Cognition

Damasio said that having a mind was the same thing as having cognition, and that even consciousness—even (conscious) *feelings*—was a cognitive phenomenon [Sects. 6.1.1.1 and 6.1.1.2]. But Panksepp, not surprisingly, takes Damasio's failure to distinguish affective and cognitive forms of consciousness to be problematic. Panksepp acknowledges that emotions and cognitions may be "quite impossible to separate unambiguously in most human enquiries" (2003b: 5), but asserts that it is still essential to study emotions separately, because "even highly cognized human emotions probably never lose the intrinsic grounding in affective primes, or else they would cease to exist" (2003a: 118).

Moreover, as I will be explaining below [Sect. 6.3.3], almost everything to do with emotions relies on regions "very low in the brain" (that is, *not* at the 'higher' level of the cerebral cortex)—so low down, in fact, that, at these levels, "the concept of 'cognition' would have to be stretched to its limits to make much sense" (Panksepp 2005: 63). But saying that, at these levels, the concept of cognition would have to be stretched to its limits to make much sense sounds very nearly like saying that it *would not* make sense to suggest these brain levels were involved in 'cognition.' Proponents of the affect-centric view assume that very little perceptual ability or cognitive sophistication is required to enable a minimal level of rudimentary affective awareness. In other words, the affect-centric view is characterized neither by an emphasis on perception nor an emphasis on cognition.

I claimed that the body-map-based view was characterized by its emphasis on perception and cognition [Sect. 6.2.1.1], which was problematic. Insofar as it is characterized neither by an emphasis on perception or an emphasis on cognition, the affect-centric view is therefore more satisfactory than the body-map-based view, from the point of view of anyone who is interested in what-it-is-like/experience.

6.3.1.2 Human Reason Or Consciousness

6.3.1.2.1 Human Reason

As I explained [Sect. 6.2.1.2], Damasio is primarily interested in 'human reason,' rather than 'human what-it-is-like/experience.' But Panksepp agrees that this is problematic, because "[a]lthough our higher cerebral functions have led to the great achievements of humankind … they have also generated *the illusory half-truth* that humans are rational creatures above all else" (1998a: 300, my emphasis).

> Although the later arriving 'higher' cognitive functions have a certain partial autonomy, any notion of full independence is illusory. One needs only to be in a life-threatening situation, struggling with terrible hunger or thirst, or in great pain to appreciate this fact. In all of these situations, homeostatic mandates dominate any and all cognitive processing and conscious focus. Watt (2005: 84)

Although much of Panksepp's recent work has been in the field of animal behavioral science, his background incorporates both neuroscience and psychology, including psychotherapy, and especially psychoanalysis—and that background shows through here. Watt's remarks about (his and Panksepp's view about) the powerful biological drives effortlessly overriding cognitive processing is very reminiscent of what Freud said [Sect. 4.3.1.2] about the instinctual (affective) urges of the id dominating all cognitive processing and conscious focus in the ego and/or superego.

6.3.1.2.2 Human Consciousness

On Damasio's 'restrictive' view, the range of creatures which can legitimately be called 'conscious' is limited to a very small segment of the spectrum: that segment where creatures have not only 'minds,' but a reflective sense of self as well [Sect. 6.1.1.1]. But as I explained, a significant number of contemporary writers think that studying something along the lines of what I am calling 'rudimentary affective awareness' is 'bound to inform us profoundly' [Sect. 5.3.1] about our own consciousness. Their view is that introspective, reflective, human consciousness is only possible in the first place insofar as a more rudimentary affective awareness underlies and supports it.

Affect-centric writers agree that to "exclude all other animals from the circle of consciousness ... makes no evolutionary sense" (Solms and Panksepp 2012: 148). They also usually agree that the main point of 'consciousness' is to provide organisms with a way of understanding the distinction between self and not-self; and all organisms have to be able to make this distinction in order to complete even the most basic of motor challenges essential to their survival (for example, Merker 2005). From this self-referenced point of view, the important questions any organism needs to be able to answer are questions like

> where am I in relation to that advancing sharp claw thing? Am I moving now, or is the environment moving? Am I eating my own arm? Asma and Greif (2012: 30)

—and a sophisticated form of reflective consciousness is not necessary for any of this. This makes it possible to say that the affect-centric view is characterized neither by an emphasis on human reason nor an emphasis on human consciousness.

In reviewing Barrett (2006), Panksepp (2007b) complained that Barrett "reviewed mainly correlational evidence" for her natural-kind model of emotions. In response, Barrett claimed that, on the contrary, in her study, "a person's mental state was manipulated and measures of subjective experience ... were taken" (Barrett et al. 2007: 299). But without reading any further than this, we can be confident that the authors are now talking at cross purposes. The whole point of the kind of 'rudimentary affective awareness' that Panksepp is interested in is that it is a *primitive* form of awareness—it provides the foundations upon which more sophisticated 'consciousness' is built, but it is not, itself, any kind of 'mental state.' This seems to be a good reason for thinking that anything Barrett has to say about mental states will constitute a misrepresentation of Panksepp's position. For Panksepp, Barrett *et al.*'s statement quoted above would carry little if any meaning (perhaps that would be like saying 'an idea's greenness was manipulated and measures of the gloves' pairhood were taken').

I claimed that the body-map-based view was characterized by its emphasis on (human/introspective) consciousness and reason [Sect. 6.2.1.2], which was problematic. Insofar as it does not attempt to account for human reason or (introspective) 'consciousness,' the affect-centric view is therefore more satisfactory than the body-map-based view, from the point of view of anyone who is interested in what-it-is-like/experience.

6.3.1.3 The Cerebral Cortex

I claimed that Damasio's view could be characterized as perceptual-cognitive [Sect. 6.2.1.1]; I said that this perceptual-cognitive view was a view of human reason and human consciousness [Sect. 6.2.1.2]; and that both these points entailed the point that Damasio also explicitly makes: his view is corticocentric. Although his later work represented some concessions towards the subcortical brain, Damasio seemed never completely to relinquish his belief that the cerebral cortex was 'the current pinnacle of neural evolution' [Sect. 6.2.1.3]. In contrast, proponents of the affect-centric view do not presume that *being the pinnacle of neural evolution* makes the *cerebral cortex* very important. They presume that *being responsible for the strongest and most primitive biological drives* makes the *subcortical brain* very important.

Panksepp (2005: 64) claims that, although cognitive (cortical) activities can suppress or modulate emotion processes in the subcortical brain, affective states can apparently only be *regulated* by cognitive/cortical activity; they apparently cannot be *created* by such activity. In contrast, when (subcortical) emotion systems are stimulated, the mind is "immediately filled with relevant, affectively-loaded thoughts" (Panksepp 2003a: 126). Although our knowledge and understanding of the subcortical brain remains disappointingly incomplete [Sect. 5.1.1.2], we are reasonably confident that "we can only provoke strong emotional feelings by manipulating brain areas below the cortex" (Panksepp 2004: 25).

Empirical studies have "commonly" shown that, when depth electrodes are placed to stimulate specific brain areas, stimulating certain subcortical regions can produce "coherent, integrated, and natural-looking ... behavioral reactions" (Merker 2007: 74), such as orienting toward stimuli, exploring the environment, and engaging in self-defensive and sexual behaviors. Stimulating any other areas, in contrast, "tends to evoke more fragmentary or incomplete behaviors" (ibid.). Moreover, all the behaviors just mentioned can also be evoked in animals which have had their cerebral cortices removed completely. Although such animals (normally rats) do show some slight differences across various dimensions of their behavior, their remaining subcortical brain structures are quite sufficient for enabling them to deal with the contingencies of day-to-day survival.

Panksepp quite unequivocally asserts that

> As long as psychology and neuroscience remain more preoccupied with the human brain's impressive cortico-cognitive systems than subcortical affective ones, our understanding of the sources of human consciousness will remain woefully incomplete. (2004, in Panksepp (2005: 31))

"Affective neuroscience reminds us of our phylogenetic homologies with other mammals, and so our biological identity should be found near the core of the brain—not the more recent neocortex" (Asma and Greif 2012: 30)—which means that not only is the affect-centric view not characterized by an emphasis on the functioning of the cerebral cortex: it is also explicitly anti-corticocentric.

I claimed that the body-map-based view was characterized by its emphasis on the functioning of the cerebral cortex, which was problematic [Sect. 6.2.1.3]. Insofar as it is not characterized by an emphasis on the functioning of the cerebral cortex, the affect-centric view is therefore more satisfactory than the body-map-based view, from the point of view of anyone who is interested in what-it-is-like/experience. In fact, in all the specific respects I have discussed in the preceding chapters, the body-map-based view is unable to avoid the pitfalls I have already pointed out, while the affect-centric view avoids them successfully.

6.3.2 Biological Value

Damasio is interested in how individuals are able to succeed in analysing complex situations, forming plans of action, or making decisions [Sect. 6.2.1.2]. The problem is that, "without the internal 'value compass' provided by emotion and affect" (Watt 2017: 184), there would be no reason for the organism to do any of that analysing, planning or decision-making in the first place. According to Panksepp, accounting for this kind of internal value compass is something that no sensory-centric view (such as the body-map-based view) can accomplish satisfactorily (2003a: 129). On Damasio's view, "we are left with a bunch of concatenated body maps" (Panksepp 2003a: 123–4) which appear to have no power to account for the fact that feelings *feel* like anything, or the fact that organisms *care* how their feelings feel.

In contrast, this is the central commitment of the affect-centric view: the essence of emotion states is not just the (sensory) *perception* of body states, but the *motivation* of the organism *to act* (in its best interests) [compare: Sect. 6.3.1.1].

> [P]ut bluntly, nature tells us in these prototype [emotional] states ... to 'get the hell out of there', 'terminate the source of the bad stuff with prejudice', or 'stick around', and get some more of the good stuff. There is simply no other way to understand what emotion really does for us and to us, no other way to see it except as the primal signature of value in the brain/mind. Panksepp (2003a: 123)

6.3.2.1 SEEKING

For Panksepp, the concept of value is epitomized by the concept of SEEKING. He identifies seven basic emotion systems as being innate to all mammals (indicating by the use of capitalization that he is referring to one of these systems, as opposed to using the word in its everyday sense); but it is fair to say that the SEEKING system comprises "a special class of one" (Watt 2005: 82), out of these seven. SEEKING can be seen as the underlying motivation for all the other emotions, because when an organism feels *fear*, for example, it is motivated to SEEK safety. When it feels *anger*, it SEEKS to terminate the circumstances causing its anger. Even *feelings of maternal care* can be conceived of in terms of the SEEKING of the safety or well-being of that vulnerable infant the mother cares for.

Taking the SEEKING system as the prototype of an emotion system has the advantage of showing how the organism's biological imperative to maintain homeostasis causes it to experience affective feelings. "[A]s Jaak famously offered, 'the 'SEEKING' system gets thirsty animals to water, hungry animals to food, cold animals to warmer environments,' etc." In this way, Panksepp's view of the centrality of the SEEKING system provides "a highly plausible theoretical bridge for how homeostasis ... might have given rise to the emotional primes" (Watt 2017: 187)—which is what views about bunches of concatenated body maps [compare: Sect. 6.3.2] are completely unable to do.

The appetitive engagement with the world embodied in the SEEKING system underlies not only our own consciousness, but also all types of motivation, for all types of action, in all types of creatures—and it has very little to do with perception/cognition, human consciousness/reason, or the cerebral cortex. It requires only the kind of rudimentary affective awareness which can be found in decorticated humans, as well as plenty of other organisms.

6.3.2.1.1 SEEKING and the Brain

When it was found (during the behaviorist era) that laboratory animals learned to seek artificial stimulation of certain areas of the brain, this seemed like a promising direction in which to search for the basis of 'value' (Panksepp 2005: 46). Insofar as animals would seek out such stimulation whenever they possibly could, it seemed that the brain systems being stimulated must be functioning as a 'reward' for those animals. However, Panksepp considers understanding these results in such simplistic terms as 'stimulus' and 'reward' to be problematic:

On the behaviorist account, the fact that an organism performs a certain action is explained by the fact that the organism has learned to associate a certain positive outcome with that action (in other words, something has 'reinforced' that behavior)—but, as Panksepp points out, "the evidence is overwhelming that *affects are the only true reinforcers*" (Watt 2005: 85). For example, a researcher might say that the rat's lever-pressing behavior is 'reinforced' if the rat gets a food pellet every time it presses the lever—which seems self-evident—but getting a food pellet can only 'reinforce' a behavior *if getting a food pellet produces a positive affect in the rat.* If getting that food pellet did not make that rat (in at least some sense) *feel good*, then getting that food pellet would not function as a 'reinforcement' of the lever-pressing behavior. In the age of information-processing, it is easy to forget this: organisms *need motivation* before they would have a reason to do any action at all—and reasons to act are provided by affective values.

On the body-map-based view, 'biological value' is seen as being produced by brainstem structures. However, even when Damasio admitted [Sect. 6.2.3.4] that the brainstem was very important in a number of different respects, it still seemed plausible to classify his view as a species of '[smart-]arousal' view: the brainstem was indispensable for virtually all biological and psychological functioning, but it was indispensable primarily (if not *exclusively*) insofar as *it enabled the cortex* to function. Much the same thing also applies to what he has to say about Value: "[t]he brain-stem component of wakefulness process relates to the natural value of each ongoing situation ... Value determines ... how awake and alert we are to be" (Damasio 2010: chap. 8, sect. 4). The brainstem helps to produce 'value,' and value, in turn, tells *the cortex* the degree to which it needs to be aroused.

Obviously, any SEEKING animal has to interact with the environment in order to acquire whatever it is seeking; and in order to interact with the environment, at least rudimentary levels of perception and cognition are required. And as we know, almost all perception and cognition rely on cortical functioning [Sects. 5.1 and 5.2]. However, on Panksepp's view, the SEEKING system itself is constituted by a neural pathway running from the midbrain to the cortex (1998a: 53–4). This system helps generate a positively-valenced psychological state which promotes resource acquisition, by creating a general "invigorated positive feeling of engagement with tasks" (Panksepp 2005: 48)—and it is located in the (subcortical) brains of (all) mammals.

Biological value cannot be satisfactorily accounted for by sensory-centric views. The affect-centric view, in contrast, not only accounts for biological value satisfactorily, but assigns biological value a central role in its explanatory framework. Therefore, the affect-centric view is more satisfactory than the body-map-based view—not only from the point of view of anyone who is interested in what-it-is-like/experience, but from *any* point of view.

6.3.2.1.2 The Affect-Centric View of Phineas Gage

On Damasio's view, Gage, after his injury, was not able to make use of 'somatic markers' during the decision-making process, and that was why he was so bad at making decisions, and generally at coping with his life [Sect. 6.2.2.1]. But I explained that 'secondary emotions' (including 'somatic markers'), on Damasio's view, necessarily included a 'cognitive' component, insofar as they involve the organism's ability to appreciate the relation between a cause and its effect. And it is not at all controversial to say that cognitive functioning is primarily dependent on the cerebral cortex. Therefore, if we discovered that Gage's (cognition-requiring) 'secondary emotions' were compromised by his (cortical) frontal lobe injury, that would be no more than we might have expected—which is why I did not consider it plausible for Damasio to say that Gage experienced deficits relating to his 'emotions.'

Panksepp agrees that Phineas Gage had difficulties with decision-making; but whereas Damasio said that the damage to Gage's orbitofrontal cortex had compromised his ability to experience 'secondary emotions,' which in turn was what compromised his decision-making, Panksepp suggests that the real problem was that Gage had been "stripped of … a life time accrual of social meanings and the contingencies for affective reward and affective aversion" (2003a: 118). The *affective reward* and *affective aversion* Panksepp is talking about, in themselves, may be paradigmatically affective in nature; but the *social meanings*, and the *contingencies* for the affective reward and aversion, are not. Meanings and contingencies can be appreciated only through the application of cognitive processing, which depends on the cerebral cortex.

The case of Phineas Gage provided some of the earliest evidence that "damage to certain brain regions (for example, the medial prefrontal cortex) interferes with social and emotional competence while not affecting cognitive competence in other domains" (Heatherton et al. 2004: 974)—but what seems significant to me here is that these authors also appear to be taking it for granted that social and emotional competence are domains of *cognitive* competence. In other words, while Damasio thinks it significant that Gage had deficits related to *social and emotional domains* (of cognition), the way these authors phrase this point shifts attention to the fact that these are both deficits in (the social and emotional domains of) *cognition*. This would be consistent with the affect-centric view that the cerebral cortex is essential for most cognition, and plays only a limited role in affective experience.

As we have seen [Sect. 6.3.1.1], Panksepp thinks that affects and cognitions are significantly distinct types of processes which should not be excessively or prematurely conflated with each other. But excessively and prematurely conflating affects and cognitions appears to be exactly what Damasio is doing in his discussion of Phineas Gage. Maybe we could put it as follows: Damasio thinks that Phineas Gage's impairment is an impairment of '(his cognitive processing of, among other things) *various emotions*'; Panksepp, we must assume, would say that Gage had an impairment of 'his *cognitive processing* (of, among other things, various emotions).'

Of course, in order to evaluate these two ways of accounting for the case very thoroughly, we would have needed to know whether Gage was able to produce skin conductance in the startle condition [compare: Sect. 6.2.2.1], and then whether, like Damasio's frontal patients, he subsequently failed to produce skin conductance when expected, when viewing the 'disturbing' images. This kind of testing was presumably not available to doctors in the mid-nineteenth century, but seeing that Dr Harlow ascribed to Gage 'the animal passions of a strong man,' I assume that his skin conductance in the neutral and startle conditions would have been normal. I also assume that the only place where he would have performed abnormally (like the other frontal patients Damasio mentioned) would have been in those trials where he failed to experience a genuine emotional reaction in a situation that would normally be expected to elicit such a response. But I have already explained why such results seem to fall short of supporting the somatic marker hypothesis—and it still seems much more plausible to me to account for Gage's deficits in purely cognitive terms (in terms of his loss of the necessary 'social meanings and contingencies'), and without referring to his 'emotions.'

Therefore, the affect-centric account of the case of Phineas Gage is more satisfactory than the body-map-based account.

6.3.2.2 The Affect-Centric View on the Evolutionary Advantages of Rudimentary Affective Awareness: Feeling Without Knowing

Descartes said that "any given movement occurring in the part of the brain that immediately affects the mind produces just one corresponding sensation; and hence the best system that could be devised is that it should produce the one sensation which, of all possible sensations, is most especially and most frequently conducive to the preservation of the healthy man" (1641/1996: 60). In other words: if the brain produces the sensation of consciousness, this fact is evidence that the sensation of consciousness is beneficial to man's evolutionary fitness.

Now isn't this exactly what we think today? Darwin proposed that "any being, if it vary however slightly in any manner profitable to itself, under the complex ... conditions of life, will have a better chance of surviving, and thus be NATURALLY SELECTED" (Darwin 1959, Introduction); and for any kind of biological naturalist, the only way anyone can explain anything about biological organisms is by referring to how a certain attribute could enhance a certain organism's survival fitness. As a matter of logic, any capacity that an organism has must in some way have provided that organism's forebears with some kind of survival advantage—otherwise that tendency would have become extinct.

Seeing that the affect-centric view is strongly committed to biological naturalism, it is not surprising to find that "in the last thirty years, affective neuroscience (AN) has successfully wedded evolution with neuroscience" (Gabriel 2012: 19). Moreover, Panksepp says that "*the evolutionary view* [is] that an emotive system for resource acquisition is an intrinsic part of the nervous system" (2005: 49, my emphasis); evidently, he is using 'the evolutionary view' to refer to his own view (which I have been calling 'affect-centric')—which clearly indicates how central evolutionary considerations must be to him.

Whereas Damasio is only prepared to ascribe 'consciousness' (that is, 'mind' + 'self') to fairly sophisticated, cognitive creatures—and maintains that 'consciousness' gives these creatures even more survival advantage than just 'mind' on its own—Panksepp thinks that a kind of rudimentary affective awareness is common to all mammals, and that this is what gives them their survival advantage. In fact, on the affect-centric view, it would probably be fair to say that the distinction between 'mind' and 'consciousness' is not even meaningful [compare: Sect. 6.1.1.1]. For Panksepp, primary phenomenal states comprise both 'the ability to perceive and orient in the world,' and, more importantly, 'the ability to feel the biological values of existence' [Sect. 5.3.1.5]; and neither of these abilities require (introspective) 'consciousness.' For Panksepp, all organisms which enjoy at least a basic level of affective awareness benefit equally, irrespective of whether they have enough cognitive sophistication to experience true 'consciousness,' in Damasio's sense. Therefore, for Panksepp, there is no important difference between 'the evolutionary advantages of mind,' and 'the evolutionary advantages of consciousness.'

6.3.2.2.1 Novel/Flexible Responses

Whereas Damasio said that 'consciousness' enabled organisms to produce novel responses to new and unfamiliar situations which they have not been designed to respond to automatically [Sect. 6.2.2.2], on the affect-centric view, organisms are thought to be "born with a variety of innate affective tendencies" (Panksepp 1998a: 24), supported by the subcortical brain structures which are common to all mammals. Organisms would therefore presumably be able to produce sensible reactions in completely new situations, not because of their advanced (cognitive) capacities for weighing a number of considerations against each other, or for generating novel responses, but because of these *innate* (unconscious, instinctual) affective tendencies.

Panksepp agrees that an important advantage associated with 'consciousness' is that it allows conscious creatures to behave more flexibly (1998a: 72). Behavioral flexibility is particularly useful in "complex environmental situations in which several behavioral alternatives are competing, with comparable urgency" (Panksepp 2001: 153), which could explain why humans' 'consciousness' has developed further than other creatures.' If consciousness enhances behavioral flexibility, and if behavioral flexibility is required for choosing between competing behavioral alternatives, then it is not surprising to find that humans—who have to deal with much more complex and multi-faceted types of environmental challenges than any other species—also enjoy a much more advanced level of 'consciousness.'

But whereas Damasio thinks that the reflective awareness of internally-displayed *images* (especially body-maps) helps 'conscious' creatures to respond more flexibly [Sect. 6.2.2.2], Panksepp thinks that sentient creatures achieve behavioral flexibility by (consciously) "dwelling on events and their meaning, as guided by *internally experienced emotional feelings*" (1998a: 38, my emphasis). Although Damasio does have the conceptual vocabulary for distinguishing between 'emotions' and 'feelings of emotion' [Sect. 6.1.1.2], he thinks that even feelings of emotion are a species of perception (of bodily processes). 'Consciously dwelling on events, as guided by internally experienced emotional feelings,' is presumably not the kind of thing Damasio ever thinks about, or is at all interested in.

6.3.2.2.2 Planning Ahead

Damasio thought that one of the advantages of 'consciousness' was that it enabled complex organisms to anticipate, and pro-actively address, their needs, before those needs became too pressing [Sect. 6.2.2.2]. Similarly, according to Panksepp, affective experience enables organisms to anticipate life-challenging situations, rather than just reacting to them. The affective feeling of hunger, for instance, is not produced only when the organism's energy supplies are completely depleted, but even when they are just starting to run a bit low. In contrast, Damasio takes the 'anticipating' that is necessary for enabling the organism to plan ahead to be largely a cognitive activity. On the affect-centric view, rudimentary affective experience helps the organism not so much with *planning* ahead, as *being motivated* (ahead of time) to take the necessary steps. Becoming aware of the affective feeling of hunger *motivates* the organism to start taking steps towards getting hold of something to eat, *to appease its feeling of hunger*, well before its blood sugar levels become critical—and it is this motivation (rather than its ability to start constructing a plan of action) which is advantageous to the survival of the organism.

6.3.2.2.3 Feelings/Incentive

Damasio thinks that ('conscious') emotional feelings are needed to give organisms enough incentive to carry out the correct survival-promoting behavior—he presumably has in mind something along the lines of 'Sometimes, cognitive evaluation of a situation does not provide sufficiently strong motivation to ensure the necessary kind of survival-promoting behavior, which is why the additional incentive of a pleasurable or unpleasurable feeling is needed.' It sounds as if Damasio is thinking of the kind of 'cognitive evaluation' of the situation one might almost call 'a cost-benefit analysis.' On his view, this analysis on its own might not be sufficient to produce a fast and strong survival-promoting reaction, and therefore it might be reasonable to suggest that some additional motivation was needed—*if cost-benefit analysis really were what was taking place in that survival-relevant situation.*

However, a good number of species are not capable of such a rational thing as a cost-benefit analysis (and even persons are often a lot less rational than they like to think [Sect. 4.3]). All creatures—even human beings—stop thinking rationally, and start reacting instinctively, as soon as they find themselves in any life-threatening situation (for example, Watt 2005, in 6.3.1.2 above); which the affect-centric view acknowledges. On this view, affective feelings are seen as beneficial not insofar as they give the organism an *incentive* to heed the results of its cost-benefit analysis, but insofar as they *effortlessly override* any kind of 'analysis.' They are not so much an *aid* to rational, cognitive (cost-benefit-analysis) processes, as an *alternative* to such processes.

For example: "Under normal circumstances the adjustment of respiratory rate and … volume needed to keep blood gases within normal bounds is automatic, effortless, and unconscious." But in circumstances where the blood gases get out of this optimal range, that fact "intrudes most forcefully on consciousness in the form of an acute sense of panic."

> Extreme blood gas values are an indication that urgent action on the environment—such as removing an airway obstruction or getting out of a carbon dioxide filled pit—may be imperative. That is [why] respiratory control … 'enters consciousness' in the form of a powerful feeling of suffocation. Merker (2007: 73)

Because efficient respiration is so vital, the organism must be strongly motivated to take action if its respiration is under threat, and its concern for its respiration "ought momentarily to take precedence over all other concerns" (Merker 2005: 97–8). That is why *the affective feeling of panic*, as opposed to the reflexive response programs of primitive organisms—*or even as opposed to the ability to plan ahead or produce novel responses* (which is what Damasio thinks is so beneficial)—is so effective in motivating creatures to promote their own survival. Damasio, as I have explained [Sect. 6.1.1.2], understands 'feelings' as a species of 'perception,' which is why he is unable to account for the urgent action impulse which is so characteristic of panic feelings (for example, Panksepp 2003a: 123–4).

Therefore, it seems safe to say that the affect-centric answer to this very important question about the evolutionary advantages of consciousness is more satisfactory than the body-map-based answer.

6.3.2.3 The Lesson (of 'Biological Value') for Philosophy

Comparing the affect-centric and the body-map-based understanding of biological value brings the differences between the views into sharper focus. For Damasio, 'biological value' is epitomized by (perceived) body states which are in a good homeostatic balance; but on the affect-centric view, the point is not that the organism has to perceive its own body states, in order to determine whether or not the thing causing those states has value to it. The point is that the organism has to be (automatically and unconsciously) *strongly motivated* to pursue whatever is of value to it because of the *overwhelming urgency* of its affective feelings.

Panksepp does not differentiate between what Damasio refers to as 'emotions' and ('conscious') '*feelings of* emotion,' because for Panksepp (as for Freud, Sect. 4.3.2.2), there is no such thing as an affect which is completely unconscious [Sect. 6.1.1.2]. Affective feelings, according to proponents of the affect-centric view, have value to organisms because they can powerfully motivate creatures to act immediately and decisively in their own best interests—and this power of motivation is not significantly affected by creatures' level of cognitive sophistication (including the capacity for reflective 'consciousness'). Whereas Damasio thinks that *being conscious* of feeling an emotion is significantly more advantageous than just *feeling* an emotion [Sect. 6.2.2.2], proponents of the affect-centric view would probably say that this capacity for 'consciousness' might be useful for things like problem-solving, but would not affect the efficacy of affective motivation much, if at all. If anything, affect might even be a *more* powerful motivator in creatures which *do not* have enough cognitive sophistication to think about their own reactions in a 'conscious' (reflective, self-aware) way.

In a sense, the way a certain framework is able to account for the evolutionary fitness of the species is the ultimate test of the view: for anybody who is not a dualist, and who embraces the Darwinian principle of evolution by natural selection, accounting for exactly what characteristics provided creatures with a survival advantage, and how, and why, is the only way of really 'explaining' anything about those creatures. Therefore, this is also the most important thing that any naturalistic account must be able to do credibly. And, as far as I can see, the body-map-based view, like other 'sensory-centric' views, is not credible on this topic. On the other hand, the affect-centric view provides a persuasive account of how being 'conscious' of affective feelings has benefitted organisms in their struggle for survival.

6.3.3 Panksepp's Views on the Subcortical Brain

The affect-centric view takes it for granted that affect is produced subcortically; but as I noted [Sect. 5.1.1.2], the subcortical brain is not very easy to visualize. The last few decades have seen the development of some very powerful new techniques which can be used for visualizing the subcortical brains of animals, but these techniques will never be applicable to human studies, since they require the complete removal of the animal's brain immediately after the completion of the task (Panksepp 1998a: 92). Moreover, even if we do manage to visualize the relevant subcortical systems, in humans or other animals, the functioning of these systems is so complex that Panksepp thinks we are still centuries away from being able to say that we really understand how they operate. Even as he outlines his views, he acknowledges that "we may find there is a great deal more variability, plasticity, and flexibility in the underlying systems" (Panksepp 1998a: 319) than even he has suggested.[9]

6.3.3.1 Brainstem and Arousal/Life Regulation

Damasio admits that nuclei in the upper brainstem not only are essential to 'life management,' but also comprise "the neural home of biological value"—which is, in all likelihood, "the place where the process of making mind begins, in the form of primordial feelings" (Damasio 2010: chap. 10, sect. 2). However, as we know, Damasio thinks that the life-regulation functions of the brainstem are important mostly insofar as they provide the necessary underpinning for the functions of *the cortex* [Sect. 6.2.3]. For instance, although the brainstem may be the home of biological value, the main important thing about biological value, for Damasio, is that it provides the cortex with the requisite level of wakefulness [see Sect. 6.3.2.1].

On the affect-centric view, in contrast, the life-regulation functions of the brainstem which are related to maintaining homeostasis in the body crucially "must inform primitive ['conscious'] qualia"—because

> [i]f consciousness does not pay very close attention to status of these fundamental homeostatic operations, it will fail quickly in its primary task of preserving the body, and clearly evolution did not select for anything that biologically foolish. Watt (2005: 84)

Whereas the body-map-based view takes 'life regulation' to be a sort of necessary, but unimportant and uninteresting, prerequisite for 'higher' cortical functioning, the affect-centric view takes life regulation to be *the most* important and interesting capacity of any living organism, and therefore the one overarching goal towards which all the organism's actions must be directed. In the same way that rudimentary affective awareness provides the foundation for all (other) forms of consciousness [Sect. 5.2], the capacity to regulate life provides the foundation for all (other) capacities.

6.3.3.2 Brainstem and Motor Maps

On Panksepp's view, the essential groundwork for the creation of 'the self' is laid down by the "stable motor coodinates" (1998a: 309) of the brainstem. But although this sounds rather like what Damasio said about body maps, the kind of motor map Panksepp is thinking of, and Damasio's kind of map of the body, are actually very different.

Whereas Damasio said that it was primarily the sensory regions of the cortex that were responsible for making body maps, Panksepp thinks that it is subcortical motor processes which underlie a basic form of 'primordial self-schema' (Panksepp refers to it as the SELF, short for 'Simple Ego-type Life Form')—the main function of which is to enable organisms to distinguish between self and non-self. For Panksepp, the "deeper layers of the colliculi" (that is, 'subcortical' structures in the midbrain) "constitute a basic motor mapping system of the body" (1998a: 312) rather than being responsible for sensory maps. These motor maps are the kind of things which might enable one, for example, to pick up one's foot, and place it in front of one's other foot, without needing first to look down and see the position of both feet.

In addition to being responsible for motor-mapping, this midbrain system also "interacts ... with nearby emotional circuits of the PAG [periaqueductal gray]" (Panksepp 1998a: 312)—and as we know, emotional circuits are central to the affect-centric view. The PAG is responsible for more "visceral-type" maps of the body ('mapping,' for example, the temperature of the body, or 'centralized' feelings such as thirst or fatigue); it is also right next to the mesencephalic locomotor region, which is heavily involved in all voluntary movement. Panksepp therefore believes that this whole midbrain region, and the motor maps it produces, are more foundational to the primordial self-schema (SELF) than the kind of sensory body-maps which are so central to Damasio (Panksepp 1998a: 313).

But it seems to me that Panksepp's views on motor maps have been the source of some confusion. According to Asma and Greif, the main difference between AN and conventional thinking in philosophy is that philosophers "tend to think about consciousness and subjectivity in terms of perceptions," whereas "affective neuroscience reminds us that 'a level of motor coherence had to exist before there would be utility for sensory guidance' (Panksepp 1998a)" (Asma and Greif 2012: 30). Presumably, what they are thinking of is the following passage:

> If one had to select between these functions (the motor or sensory zones of the tectum) as the very focus of the SELF process, I am inclined to envision the motor map as being more central to the SELF than the incoming somatosensory processes. This is based partly on evolutionary considerations: A level of motor coherence had to exist before there would be utility for sensory guidance. Panksepp (1998a: 312)

So, as we can see, Panksepp did say that a level of motor coherence had to exist before there would be any point in having sensory maps—but what he also said was that '*if one had to select* between motor or sensory functions' as being foundational for the self, he would be *inclined* to favour motor maps as being *more central than* sensory maps (see also Panksepp 1998a: 567). Seeing that I have already explained at some length how Panksepp's view is anti-sensory-centric [Sect. 6.3.2], finding that he sees motor maps as *more* central to the construction of the SELF *than sensory maps* is not much of a surprise. Moreover, saying that 'A level of motor coherence had to exist before there would be utility for sensory guidance' should also not come as much of a surprise: it would have been necessary for ancestral organisms to be able to move, and to map their own movements in some rudimentary way, long before the organisms became sufficiently sophisticated to need the kind of sensory perception that would have enabled them to construct any kind of (sensory) body maps.

For Damasio, these regions of the subcortical brain are important for enabling the organism to *make [coarse] maps* [Sect. 6.2.3.3]; for Panksepp, they are essential to the organism's ability to perceive *and orient* in the world [Sect. 5.3.1.5]. The ability to perceive-and-orient is not the kind of capacity that relies too much on sensory perception, or even maps of one's own body. All it requires is a very rudimentary sense of oneself as an embodied being in the physical world. The perceive-and-orient maps of the brainstem are not the kind of maps that enable somatic markers to act as beacons of incentive or alarm bells to those engaged in complex decision-making [Sect. 6.2.2.1]: they are the kind of maps that enable organisms to know whether they, or their environments, are moving; to identify their own positions in relation to advancing sharp claw things; and to avoid eating their own arms [compare: Sect. 6.3.1.2].

Moreover, these kinds of basic motor maps might be essential for perceiving and orienting in the world—but we must remember that, for Panksepp, the component of primary phenomenal experience which involves the ability *to perceive and orient in the world* was not as important as the second component of primary phenomenal experience: being able *to feel the biological values of existence* [Sect. 5.3.1.5]. Perceiving-and-orienting requires motor maps, but feeling the biological values of existence does not.

6.3.3.3 Brainstem and Feelings

In a sense, it is redundant to have a specific section for discussing Panksepp's views on the brainstem and feelings, since the affect-centric view might well be classified as a view *about* the importance of feelings, which entails that it is also (therefore) a view about the importance of the brainstem/subcortical brain. But whereas Damasio was concerned with emotion feelings largely insofar as they disrupted or enhanced reasoning and decision-making, Panksepp emphasizes the extent to which the intrinsic motivational urgency of affect *overrides* any kind of reasoning process [compare: Sect. 6.3.2.2]. Moreover, on the affect-centric view, emotional feelings are what enable us to have higher cognitive abilities (like reasoning and decision-making) in the first place [compare: Sect. 5.3].

Affective states cannot be generated by purely cognitive activity, whereas affective states influence cognitive processes heavily; stimulating certain subcortical regions produces reasonably coherent behavior patterns, whereas stimulating cortical regions does not; and even completely decorticated animals are competent to deal with most contingencies [Sect. 6.3.1.3], whereas even partial damage to the brainstem usually results in the loss of consciousness [Sect. 5.2]. On the affect-centric view, basic (subcortical) affective feelings provide the essential foundation both for all types of what-it-is-like/experience, and for all types of 'higher' mental capacities, such as (reflective/introspective) 'consciousness' and rationality [Sect. 5.3].

6.3.3.4 Brainstem and Conscious Minds

It hardly seems controversial to claim that primitive ('conscious'—because there is no such thing as an affect which is not felt, Sect. 6.1.1.2) affective states are shared by all mammals, but "this basic proposition is still spurned in many sectors of behavioral neuroscience as anthropomorphism" (Watt 2005: 85). In fact, because behaviorism still has such a surprising amount of influence [Sects. 4.1 and 6.1], saying that this basic proposition is spurned *in many sectors of behavioral neuroscience* may be tantamount to saying that it is spurned *tout court.*

We already know that Damasio, so far from agreeing that affective consciousness provides an essential foundation for complex cognitive consciousness, is not really even able to *recognize* the concept of 'affective consciousness': he *defines* 'consciousness' as comprising both 'mind' and 'an automatic sense of self as protagonist' [Sect. 6.1.1.1], which means that, for him, more rudimentary forms of affective awareness do not count as 'consciousness'; moreover, for him, consciousness is a requirement for (fully) experiencing feelings [Sect. 6.1.1.2]. In contrast, the affect-centric view is that basic emotional affects, together with basic sensory affects (both of which can be generated subcortically), "form the basis for more affective forms of consciousness that in turn are the essential foundations for more complex cognitive consciousness" (Watt 2005: 88) [compare: Sect. 5.3]. And these more affective forms of consciousness (which are essential for more complex cognitive consciousness) are almost entirely independent of the functioning of the cerebral cortex.

6.3.3.5 The Lesson (of 'Panksepp's views on the subcortical brain') for Philosophy

On the affect-centric view, the subcortical brain provides the essential foundations for affective consciousness, which in turn provides the essential foundations for all types of 'higher'/cognitive capacity; including, ultimately, the capacity for (introspective) 'consciousness.' But Panksepp cautions that affect, and its role in the development of 'higher' human capacities, is such a complex topic that one should be wary about making any categorical claims about it; he even concedes that it is possible that the activity of cortical (cognitive) brain systems could constrain the subcortical emotion systems he identifies, rather than the other way around. However, he reminds us that "it is also important to emphasize how much affect can still be elaborated in the human brain" (2004: 59) when most of the cerebral cortex is completely absent (as in hydranencephaly, 4.3). The fact that the affect-centric view is anti-corticocentric makes it more satisfactory than the body-map-based view.

6.3.4 The Affect-Centric View: Conclusion

In Sect. 6.3, I have shown that Panksepp's affect-centric view is far less problematic than Damasio's body-map-based view, with regard to the five specific attributes of both views which I have discussed.

Damasio's view revolved around perception/cognition, human reason/human consciousness, and the cerebral cortex, which I claimed was problematic. I have now explained in detail how the affect-centric view does not emphasize perception/cognition, human consciousness/rationality, and/or/therefore the cerebral cortex. Insofar as the affect-centric view does not share the (problematic) central commitments of the body-map-based view, the affect-centric view is more satisfactory.

The inability of the body-map-based view to account satisfactorily for biological value strikes a devastating blow to its credibility. I have now explained how the affect-centric view is able to provide a far more satisfactory account of biological value (centred around the subcortical SEEKING system) than is the body-map-based view. I have shown how the affect-centric view of the evolutionary advantages of rudimentary affective awareness is more satisfactory than the body-map-based view of the evolutionary advantages of consciousness; and I have even shown how the affect-centric view makes better sense of the case of Phineas Gage than does the body-map-based view.

Because the subcortical brain plays such a critical role, on the affect-centric view, understanding how the role of the subcortical is conceptualized helps deepen our understanding of the essential commitments of this view; and especially of how these essential commitments contrast with the commitments of the (sensory-centric) body-map-based view propounded by Damasio.

To the extent that the commitments of the body-map-based view are problematic, and to the extent that the affect-centric view rejects these commitments, the affect-centric view should be preferred by anyone who is interested in what-it-is-like/experience. We *should* use a biological naturalist framework—something along the lines of the affect-centric view—to enable us to understand the mind-body-self. Just as Freud attempted (but unfortunately failed) to account for psychological phenomena in naturalistic, biological terms, so the affect-centric view attempts (more successfully, one hopes) to understand the most advanced human consciousness phenomena in terms of basic biological drives of the subcortical brain.

6.4 Biological Naturalism: The Body-Map-Based View and The Affect-Centric View: Conclusion

The reason I wanted to do any of this research in the first place was that I was not satisfied with any of the (philosophical) accounts of the mind-body-self I had come across. In Chap. 2, I discovered that the main reason for this was that anybody who talks about 'the mind' or 'consciousness' as though it 'exists' (that is, as if it is different from physical reality) must be a dualist, whatever s/he says to the contrary; I also discovered how pervasive the false dichotomy of 'mind' and 'body' is among philosophers—even those who profess to be naturalists. But for the purposes of my argument, I am assuming that dualism is no longer a realistic worldview for the mind sciences of the twenty-first century.

In Chap. 3, I showed that, because both perception and cognition are independent of 'consciousness,' we should be careful not to conflate *consciousness* with (conscious) *perception* or *cognition*. This also entailed that we should be careful not to conflate *what-it-is-like/experience* with (perceptual-cognitive) *'consciousness'*. In Chaps. 4 and 5, I hope I have argued convincingly that we should be careful not to conflate (psychological) what-it-is-like/experience with (introspective) consciousness or rationality—and that, therefore, we should avoid limiting our investigations exclusively to the functioning of the cerebral cortex. In Chap. 6, I have now shown that:

(1) Although Damasio's body-map-based view is naturalistic—and does not qualify as an example of even 'closet' dualism—it is still characterized by its emphasis on perception, cognition, human reason and human consciousness (and therefore the functioning of the cerebral cortex). Therefore, this view is problematic from the point of view of anyone who is interested in what-it-is-like/experience [Sect. 6.2].

On the other hand,

(2) To the extent that the commitments of the body-map-based view are problematic, and to the extent that the affect-centric view rejects these commitments, the affect-centric view should be preferred by anyone who is interested in what-it-is-like/experience [Sect. 6.3]. We *should* use a biological naturalist framework—something along the lines of the affect-centric view—to enable us to understand the mind-body-self.

Notes

1. Papineau (2020) understands 'naturalism' very broadly, incorporating all views according to which "reality is exhausted by nature, containing nothing 'supernatural'"—which is the understanding I will use in this chapter.
2. Portions of this chapter (especially Sects. 6.1, 6.2.2.2, and 6.3.2.2) are reproduced in Anderson (2019).
3. Damasio also uses 'cognition' synonymously with 'mental states' (1999: 85).
4. For this reason, one should be cautious of assuming that Damasio has provided a superficial or incomplete treatment of some topic, when he may just be trying to simplify a complex issue so that it is easier for general readers to appreciate (for example, Panksepp 2003a: 118–9).
5. Or 'life regulation,' see Parvizi and Damasio (2001: 136).
6. I think Panksepp is making a similar point when he says that, in *Looking for Spinoza*, Damasio is still phrasing his claim about the role of the brainstem in a "slightly passive" way: subcortical structures "'monitor' and 'map' the state of the body" (Panksepp 2003a: 125)—as opposed, for example, to taking an active part in *causing* the state of the body.
7. Insofar as Damasio does not have the same interest in what-it-is-like/experience that I have, one could perhaps say that my remarks do not function as criticisms of his actual view—just as I said I was not trying to assess those four philosophers' views comprehensively (compare: beginning of Sect. 6.2.3). My claim is that Damasio's framework does not help us *to understand what-it-is-like/experience*, and *not* that Damasio's framework is *incorrect.*
8. Whereas I have reserved the term 'body-map-based' to refer exclusively to Damasio's views, in this section, I will discuss several other writers whose views I consider to be 'affect-centric.'
9. In light of these kinds of considerations, the confidence with which Damasio sets out his findings may be unwarranted. Certainly, he always seemed to give the impression that he and his colleagues knew exactly what was going on [Sects. 6.2.2 and 6.2.3]—whereas, if Panksepp is to be believed, it is actually not even possible for anybody to 'know' many of those things.

References

Anderson, J. (2019). Damasio's body-map-based view, Panksepp's affect-centric view, and the evolutionary advantages of consciousness. *South African Journal of Philosophy*, 38(4): 419–432. https://doi.org/10.1080/02580136.2019.1697569

Asma, S.T., and T. Greif. 2012. Affective neuroscience and the philosophy of self. *Journal of Consciousness Studies* 19 (3-4): 26–36.

Barrett, L.F. 2006. Are emotions natural kinds? *Perspectives on psychological science* 1 (1): 28–58.

Barrett, L.F., Lindquist, K.A., Bliss-Moreau, E., Duncan, S., Gendron, M., Mize, J. & Brennan, L. (2007). Of mice and men: natural kinds of emotions in the mammalian brain? A response to Panksepp and Izard. *Perspectives on psychological science*, 2(3): 297–312. Available from Academic Search Complete database: http://0-web.a.ebscohost.com.ujlink.uj.ac.za/ehost/pdfviewer/pdfviewer?vid=5&sid=a14ce3ec-025f-4d40-b8fa-4e1af565fe5d%40sessionmgr4009

Cacioppo, J.T., and G.G. Berntson. 2004. Social neuroscience. In *The cognitive neurosciences*, M. Gazzaniga, 3rd ed., 977–985. ProQuest Ebook Central database: https://0-ebookcentral-proquest-com.ujlink.uj.ac.za/lib/ujlink-ebooks/detail.action?docID=3338826.

Chalmers, D.J. 1996. *The conscious mind: In search of a fundamental theory.* https://pdfs.semanticscholar.org/5491/3945b27ca973968174ad521aa85620f0b32f.pdf. Accessed 7 August 2016.

Damasio, A. 1994. *Descartes' error: Emotion, reason and the human brain.* New York: Avon Books.

Damasio, A.R. 1996. The somatic marker hypothesis and the possible functions of the prefrontal cortex. *Philosophical Transactions of the Royal Society of London. Series B: Biological Sciences* 351: 1413–1420.

———. 1998. Investigating the biology of consciousness. *Philosophical Transactions of the Royal Society of London – Series B: Biological Sciences* 353: 1879–1882.

Damasio, A. 1999. *The feeling of what happens: Body and emotion in the making of consciousness.* Available from Antonio Damasio https://www.scribd.com/doc/242107678/Damasio-The-Feeling-of-What-Happens-1999-pdf.

Damasio, A.R. 2003. *Looking for Spinoza: Joy, sorrow, and the feeling brain.* London: William Heineman.

Damasio, A. 2010. *Self comes to mind: Constructing the conscious brain.* New York: Pantheon Books.

Damasio, A.R., Grabowski, T. J., Bechara, A., Damasio, H., Ponto, L. L. B., Parvizi, J. and Hichwa, R. D. 2000. Subcortical and cortical brain activity during the feeling of self-generated emotions. *Nature Neuroscience* 3(10):1049–1056. https://doi.org/https://doi.org/10.1038/79871.

Darwin, C. 1959. *The origin of species by means of natural selection* (6th ed.). https://www.andrew.cmu.edu/user/jksadegh/A%20Good%20Atheist%20Secularist%20Skeptical%20Book%20Collection/Charles%20Darwin%20-%20The%20Origin%20of%20Species%20-%206th%20Edition.pdf.

Descartes, R. [1641] 1996. *Meditations on first philosophy*. In J. Cottingham, trans.-ed. https://doi.org/10.1017/9781107416277.

Gabriel, R. 2012. Modularity in cognitive psychology and affective neuroscience. *Journal of Consciousness Studies* 19 (3-4): 19–25.

Heatherton, T.F., E.A. Phelps, and J.E. Ledoux. 2004. Emotional and social neuroscience: Introduction. In *The cognitive neurosciences*, M. Gazzaniga, 3rd ed., 973–976. ProQuest Ebook Central database: https://0-ebookcentral-proquest-com.ujlink.uj.ac.za/lib/ujlink-ebooks/detail.action?docID=3338826.

Merker, B. 2005. The liability of mobility: A selection pressure for the transition to consciousness in animal evolution. *Consciousness and Cognition* 14: 89–114. https://doi.org/10.1016/S1053-8100(03)00002-3.

———. 2007. Consciousness without a cerebral cortex: A challenge for neuroscience. *Behavioral and brain sciences* 30: 63–81. https://doi.org/10.1017/S0140525X07000891.

Montemayor, C., and Haladjian, H.H. 2015. *Consciousness, attention, and conscious attention*. ProQuest Ebook Central database: https://0-ebookcentral-proquest-com.ujlink.uj.ac.za.

Ostow, M. (2004). Foreword. In *Textbook of biological psychiatry*: xv-xvii. Edited by Panksepp, J. Available from ProQuest Ebook Central database: https://0-ebookcentral-proquest-com.ujlink.uj.ac.za/lib/ujlink-ebooks/reader.action?docID=469784

Panksepp, J. (1998a). *Affective neuroscience: the foundations of human and animal emotions*. Available from ProQuest Ebook Central database: https://0-ebookcentral-proquest-com.ujlink.uj.ac.za/lib/ujlinkebooks/detail.action?docID=431176#.

Panksepp, J. (1998b). The periconscious substrates of consciousness: affective states and the evolutionary origins of the self. *Journal of Consciousness Studies*, 5(5-6): 566–82. Available from: http://www.imprint-academic.com/jcs.

———. 2001. The neuro-evolutionary cusp between emotions and cognitions: Implications for understanding consciousness and the emergence of a unified mind science. *Evolution and Cognition* 7 (2): 141–163.
———. 2003a. Damasio's error? *Consciousness & Emotion* 4 (1): 111–134.
———. 2003b. At the interface of the affective, behavioral, and cognitive neurosciences: Decoding the emotional feelings of the brain. *Brain and Cognition* 52: 4–14. https://doi.org/10.1016/S0278-2626(03)00003-4.
———. 2004. Biological psychiatry sketched – Past, present, and future. In *Textbook of biological psychiatry*, J. Panksepp, 3–32. ProQuest Ebook Central database: https://0-ebookcentral-proquest-com.ujlink.uj.ac.za/lib/ujlink-ebooks/reader.action?docID=469784.
———. 2005. Affective consciousness: Core emotional feelings in animals and humans. *Consciousness and Cognition* 14: 30–80. https://doi.org/10.1016/j.concog.2004.10.004.
———. 2007a. Emotional feelings originate below the neocortex: Toward a neurobiology of the soul. *Behavioral and Brain Sciences* 30 (1): 101–103. https://doi.org/10.1017/S0140525X07001094.
———. 2007b. Neurologizing the psychology of affects: How appraisal-based constructivism and basic emotion theory can coexist. *Perspectives on Psychological Science* 2 (3): 281–296.
Papineau, D. 2020. Naturalism. In *The Stanford Encyclopedia of philosophy*, ed. E.N. Zalta. https://plato.stanford.edu/archives/sum2020/entries/naturalism/. Accessed 19 January 2020.
Parvizi, J., and A. Damasio. 2001. Consciousness and the brainstem. *Cognition* 79: 135–159.
Solms, M., and J. Panksepp. 2012. The "id" knows more than the "ego" admits: Neuropsychoanalytic and primal consciousness perspectives on the interface between affective and cognitive neuroscience. *Brain Sciences* 2: 147–175.
Watt, D.F. 2005. Panksepp's common sense view of affective neuroscience is not the commonsense view in large areas of neuroscience. *Consciousness and Cognition* 14: 81–88. https://doi.org/10.1016/j.concog.2005.01.003.
———. 2017. Reflections on the neuroscientific legacy of Jaak Panksepp (1943–2017). *Neuropsychoanalysis* 19 (2): 183–198. https://doi.org/10.1080/15294145.2017.1376549.

7

Final Thoughts

Although I have emphasized the respects in which the body-map-based view differs from the affect-centric view (and in which the affect-centric view should therefore be preferred), it is important to note that, with regard to 'the explanatory gap,' these views are more similar than they are different. Both Damasio and Panksepp quite unequivocally locate what-it-is-like/experience within "the unified life of the spatio-temporally located organism" (Asma and Greif 2012: 33); both views reject the idea that any kind of super natural force or substance underlies what-it-is-like/experience; and both views are good examples of what Searle referred to as 'biological naturalism.

7.1 The Explanatory Gap

In Chap. 2, I determined, firstly, that it was possible to produce neither a coherent metaphysical account nor a coherent conceptual (philosophical) account of the explanatory gap. Secondly, I determined that many of the people writing about 'philosophy of mind' were either overt dualists, or closet dualists; therefore, seeing that I had dismissed dualism as unsatisfactory, I said that it was not possible to produce a coherent conceptual

© The Author(s), under exclusive license to Springer Nature Switzerland AG 2022
J. Anderson, *Biological Naturalism and the Mind-Body Problem*,
https://doi.org/10.1007/978-3-030-99684-0_7

(philosophical) understanding of the explanatory gap. Biological naturalist views reject the normal philosophical assumption that it is impossible to reconcile the belief that what-it-is-like/experience *is* something more than just a figment of our imagination with the belief that everything that exists is physical.

I said that Ryle's idea of 'category mistakes' was to a large extent coextensive with my idea of vicious reification, and to that extent, I endorsed his view. However, I complained that, having diagnosed the conceptual confusion inherent in our (reified) way of talking about the mind/body, Ryle failed to produce much in the way of an alternative conceptual account. But whereas Ryle does not provide any positive ideas about how to talk about 'mental conduct' without making a category mistake, biological naturalists very sensibly do not talk about 'mental' things ('conduct,' or anything else) at all.

> Such 'empty categories' … are even easier to create in the psychological sciences … In any event, serious investigators of the brain are loathe to contribute more verbal rubbish to existing confusions so they are prone to remain silent on such matters. (Panksepp 1998a: 305)

Damasio is fully committed to the naturalist manifesto; but to some extent he does still talk about 'human consciousness' and 'the body.' The fact that even this serious naturalist finds it difficult to avoid using terms like 'consciousness' shows how right Searle was when he claimed it was almost impossible to talk about 'mental conduct' without having recourse to obsolete Cartesian categories. The affect-centric view is more successful in breaking free of such categories, because it focuses on how certain subcortical neural circuits give rise to the kind of *basic emotions* which form the core of all ('conscious') what-it-is-like/experience.

7.2 The Hard and the Easy Problems of Consciousness

In Chap. 3, I discovered, firstly, that we should be careful not to conflate *consciousness* with ('conscious') perception or cognition—seeing that so much perception and cognition is actually unconscious. Secondly, I found that, because perception and cognition seem to be almost entirely independent of consciousness—and because there is nothing-it-is-like to experience 'unconscious' perceptual and/or cognitive processes—we should be careful not to conflate (hard-problem) *what-it-is-like/experience* with (easy-problem/perceptual-cognitive) 'consciousness'. Like most 'consciousness' researchers, Damasio tends to focus on (easy-problem) perceptual-cognitive (and/or introspective) consciousness, rather than (hard-problem) what-it-is-like/experience—especially given that he understands feelings of emotion as a species of *perception* of the body. In contrast, the affect-centric view recognizes that investigating 'conscious' perception or cognition is unlikely to tell us anything very interesting about organisms. Instead, Panksepp explicitly focuses on hard-problem 'consciousness': the experiential-feeling aspect of awareness that is very difficult to describe verbally.

7.3 (Un)consciousness, (Ir)rationality, and Psychological Experience

In Chap. 4, I explored Freud's views on the mind-body-self—especially the respect in which Freud saw human persons neither as essentially (introspectively) 'consciousness' nor essentially 'rational.' I found Freud's views very persuasive; therefore, I cautioned, we should be careful to conflate (psychological) what-it-is-like/experience neither with 'consciousness' nor with rationality. Like most 'consciousness' researchers, Damasio tends to focus on (introspective) consciousness (and/or rationality) rather than rudimentary affective experience. But insofar as Freud's view (on which consciousness and rationality are very far from being central attributes of the psychological subject) provides an excellent means of accounting for unconscious and irrational psychological phenomena, this detracts from the plausibility of Damasio's view.

Freud said that distinguishing 'the psychical' from 'what is conscious' enabled him to maintain that psychical processes were similar to any other natural (biological) processes; similarly, on the affect-centric view, (psychological) what-it-is-like/experience is accounted for in terms of basic emotional networks which are "intrinsically biological *and* intrinsically psychological" (Gabriel 2012: 25, my emphasis). Rather than investigating '(introspective) consciousness' and 'rationality,' the affect-centric view, like the psychodynamic view, is almost exclusively concerned with the (rudimentary) affective, non-introspective kind of 'consciousness'; it is concerned with 'rationality' very little, if at all.

7.4 The Brain and the Mind-body-self

In Chap. 5, I suggested (firstly) that we should not take the mind-body-self to be dependent on the functioning of the cerebral cortex; and (secondly) that we *should* take the mind-body-self to be essentially affective. Like most 'consciousness' researchers, Damasio focuses on the role played by the cortex in (introspective) consciousness and rationality. But the fact that affective consciousness does not require the cerebral cortex threatens Damasio's view. In contrast, the type of rudimentary what-it-is-like/experience that can be produced even in the complete absence of cerebral cortex is at the heart of the affect-centric view. Both the Affect-centric View and the Body-map-based View correctly emphasize the importance of affects (although they conceptualize affects in rather different ways).

7.5 Twenty-First-Century Biological Naturalism: The Body-Map-Based View and the Affect-Centric View

In Chap. 6, I claimed that, in all the respects that have been important to me in this book, the body-map-based view was problematic. Because Damasio emphasizes perception/cognition, (introspective) consciousness/rationality, and (therefore) the cerebral cortex, so strongly in his account—and because I have already explained in detail why that is inadvisable—we are more likely to come up with a satisfactory account of the mind-body-self if we situate this account within something like the affect-centric framework than if we situate it within a body-map-based-view-type framework.

Panksepp never had occasion to discuss perception/cognition, consciousness/rationality, or the cerebral cortex, specifically, as I did; however, it is easy to see how the affect-centric view performs in terms of the topics I have discussed in the preceding chapters. Therefore, although Damasio's view and Panksepp's view are both examples of 'biological naturalist' views, thinking in terms of basic affective experience, rather than of perceptual-cognitive or introspective 'consciousness' (or rationality), seems more likely to help us make sense of (psychological) what-it-is-like/experience.

Perhaps we could say that Panksepp has been able to finish the task, started by Freud in the late nineteenth century, of explaining exactly how psychical processes are similar in kind to all other natural processes. According to Hopkins (2015: 104), "[i]n his early attempts at a neuroscience of motivation and emotion, Freud described the basic mechanisms of motivation as rooted in 'endogenous stimuli' that 'gave rise to the major needs … hunger, respiration, sexuality' (Freud 1950a [1985], p. 297)"—just as Panksepp does.

On the affect-centric view, focusing on the subcortical mechanisms of rudimentary affective awareness provides a way of bridging the gap between 'the body' and 'the mind,' because it takes the question of (psychological) what-it-is-like/experience to be a question of rudimentary affective awareness, rather than a question of 'consciousness' (or perception, or cognition, or rationality).

7.6 Dualism, Biological Naturalism, and *dissolving* the Explanatory Gap

Damasio thinks that, until recently, "the mind-body problem remained a philosophical topic, outside the realm of empirical science"—that it was only since about the 1990s that "the problem finally entered the scientific agenda" (Damasio 2003: 184). He thinks that "[i]n the eyes of many scientists and philosophers, the problem is either false, or has been resolved"—but the fact that so many intelligent and educated people continue to disagree so strongly about these issues seems to indicate that "the solution either is not satisfactory or has not been satisfactorily presented" (Damasio 2003: 183).

As I explained, Damasio distinguishes between 'emotions' and *feeling* of emotion. But this is problematic for those who are not dualists. Feelings, according to Damasio, are the 'mental aspect' of neural states (Damasio 2010: chap. 4, sect. 4); they constitute "the mental level" of biological operations; and in virtue of having the necessary 'mental level,' feelings can (therefore) "enter the mind fray and influence the [biological] operations" (Damasio 2003: 177)! Given that I have argued [Sect. 2.2.3] that talking about 'mental' things and 'biological' things as if they were different kinds of things is constitutive of (closet) dualism, it almost looks as if Damasio is falling into this same (dualist) trap—but I think it is possible to acquit him of actual closet dualism:

Panksepp observes that, since Damasio is "painting a portrait of complexity with a broad brush" (Panksepp 2003a: 119), he may find it expedient to express his views in something less than a strictly accurate way, for the benefit of the "general readers" at whom his popular books are primarily directed (footnote 37, section 6.2). In other words, Damasio may not intend such terms as 'mental level' to be understood in a completely literal sense. In this case, such remarks could still be compatible with a truly naturalistic view. However, the fact that he uses these terms in an imprecise manner "leaves those who wish to deal with such issues in more explicit scientific ways perplexed as to where Damasio really stands" (Panksepp 2003a: 119).

On the other hand, there can be no doubt at all as to where Panksepp really stands on the question of (closet) dualism. According to Panksepp, none of the forms of dualism currently "enjoy the respect of serious investigators" (1998a: 59); however, he claims, "the assumption that we can adequately understand psychological processes through the mere use of verbal symbols … without relying on brain research" (1998a: 59–60)—which is still widely encountered in contemporary psychological discourse, see Chaps. 2 and 3—indicates a (concealed) form of dualism.

It seems that he believes, as do I, that anyone who uses the terms 'mind' and 'body' as if they referred to distinct entities must be a (or at least, a closet-) dualist.[1] If this is indeed the case, it appears we must agree with Panksepp that "dualism is by no means dead in philosophical or neuroscientific discussions of brain functions" (Panksepp 1998a: 337):

> Although most brain scientists are happy to believe that mental events simply reflect the neurodynamic informational exchanges within many interacting networks of the brain, and nothing else, there is still substantial controversy among philosophically-oriented investigators. (Panksepp 1998a: 336)

On the affect-centric view, in contrast, "there is no unbridgeable chasm between the nature of subjective experience and relevant brain and behavioral facts" (Panksepp 1998a: 305). Affective states are inbuilt cues that tell organisms whether to approach or withdraw from the things causing those feelings. Generating these positive and negative affective states is "a natural neurobiological function of the brain" (Panksepp 2003b: 11)—so if we take these positive and negative affective states as the basis for all types of what-it-is-like/experience, "we may gradually outgrow the illusory sense that we are creatures of two distinct realms, of mind and matter, and come to monistically accept that we are simply ultracomplex creatures of the world" (Panksepp 1998a: 302).

So long as we are attempting to account for how 'mental states' relate to 'physical states,' we necessarily fall foul of the dualism/physicalism dichotomy. If everything that exists is physical, then we must give up our belief in 'mental' things; and if some things that exist are not physical, then we must give up our belief in the principle of causal closure. But as soon as we understand 'the gap question' in terms of *feelings* (which are 'conscious' by definition), the problem goes away. Tye (1999) was purportedly talking about 'phenomenal consciousness'—that was even the title of his paper—but he inadvertently gave the lie to his supposed view that 'the gap question' was a question about phenomenal *consciousness*, when he formulated the question as being about 'why so-and-so brain processes *feel* any way at all.' Without noticing it, he was implicitly recognizing that 'consciousness' should be understood as a question about *feeling*.

Neither dualism nor materialism can make any sense of the explanatory gap between 'the mind' and 'the body'—but (especially, affect-centric) biological naturalism *can* account for ('conscious') *feeling* in terms of the functioning of physical structures. In this respect, both the affect-centric view and the body-map-based view are preferable to non-naturalist views—but since the body-map-based view exhibits those five characteristics which I have said are problematic, we should prefer something along the lines of the affect-centric view. Whereas Descartes erroneously determined that he was essentially a thing which thinks, Panksepp has a much more convincing argument for proposing that he is essentially a thing which *feels*—as is any other biological organism.

Biological naturalism (especially in the form of Affective Neuroscience) provides a legitimate way of *dissolving*, rather than *closing*, the (illusion of the) explanatory gap. I have argued that naturalist philosophers should not only enrich their theories by drawing from depth psychological views such as Freudian psychodynamics: they should also learn from (empirical) biological naturalist researchers such as Damasio and Panksepp. They should recognize that there is no such thing as 'the mind' (and/or/therefore as 'the explanatory gap'), give up on all types of dualism, and just think about how affective feelings are what causes there to be something-it-is-like, which is what provides (affectively-) conscious creatures with an evolutionary advantage over their competitors.

Note

1. In fact, Panksepp goes so far as to use 'the mind-body problem' and 'dualism' *synonymously* (Panksepp 1998a: 336).

References

Asma, S.T. & Greif, T. 2012. Affective neuroscience and the philosophy of self. *Journal of Consciousness Studies* 19 (3–4): 26–36. https://www.psychologytoday.com/sites/default/files/attachments/109303/jcs-articlefinal.pdf.

Damasio, A.R. 2003. *Looking for Spinoza: joy, sorrow, and the feeling brain.* London: William Heineman. https://ahandfulofleaves.files.wordpress.com/2013/07/looking-for-spinoza_damasio.pdf.

———. 2010. *Self comes to Mind: Constructing the Conscious Brain.* New York: Pantheon books. https://ahandfulofleaves.files.wordpress.com/2013/07/self-comes-to-mind_damasio.pdf.

Freud, S. 1950a[1985]. *A Project for a Scientific Psychology*. S.E., 1: 283–397. London: Hogarth.

Gabriel, R. 2012. Modularity in cognitive psychology and affective neuroscience. *Journal of Consciousness Studies* 19 (3–4): 19–25. https://www.psychologytoday.com/sites/default/files/attachments/109303/jcs-articlefinal.pdf.

Hopkins, J. 2015. The significance of consilience: psychoanalysis, attachment, neuroscience, and evolution. In *Psychoanalysis & Philosophy of Mind: Unconscious Mentality in the 21st Century*, 1–46. Edited by Boag, S. Available from Proquest Ebook Central database: https://0-ebookcentral-proquest-com.ujlink.uj.ac.za/lib/ujlinkebooks/detail.action?docID=2082961&query=psychoanalysis+and+philosophy+of+mind.

Panksepp, J. 1998a. *Affective neuroscience: the foundations of human and animal emotions*. Available from ProQuest Ebook Central database: https://0-ebookcentral-proquest-com.ujlink.uj.ac.za/lib/ujlink-ebooks/detail.action?docID=431176#.

———. 2003a. Damasio's error? *Consciousness & Emotion* 4 (1): 111–134. Available from Jaak Panskepp (https://www.academia.edu/17183942/Damasios_Error)

———. 2003b. At the interface of the affective, behavioral, and cognitive neurosciences: decoding the emotional feelings of the brain. *Brain and Cognition* 52: 4–14. https://doi.org/10.1016/S0278-2626(03)00003-4.

Tye, M. 1999. Phenomenal consciousness: The explanatory gap as a cognitive illusion. *Mind* 108 (432): 705–725. Available from JStor database: https://www.jstor.org/stable/2660075.

Index[1]

[1] Note: Page numbers followed by 'n' refer to notes.

© The Author(s), under exclusive license to Springer Nature Switzerland AG 2022
J. Anderson, *Biological Naturalism and the Mind-Body Problem*,
https://doi.org/10.1007/978-3-030-99684-0

FSC
www.fsc.org
MIX
Papier | Fördert
gute Waldnutzung
FSC® C083411

Kolibri 360 GmbH
Ferdinand-Jühlke-Straße 7
99095 Erfurt, Deutschland
produktsicherheit@kolibri360.de